PERMUTATIONS

PERMUTATIONS
READINGS IN SCIENCE AND LITERATURE ◆ ◆ ◆

EDITED BY JOAN DIGBY AND BOB BRIER ◆ ◆ ◆

Collages by John Digby

WILLIAM MORROW AND COMPANY *New York*

Library of Congress Cataloging in Publication Data

Main entry under title:

Permutations : readings in science and literature.

    Includes index.
    1. Science—Addresses, essays, lectures.
2. Science—Philosophy—Addresses, essays,
lectures. I. Digby, Joan. II. Brier, Bob.
Q171.P37 1985b   500   85-307
ISBN 0-688-01944-7
ISBN 0-688-01945-5 (pbk.)

Printed in the United States of America

First Edition

1 2 3 4 5 6 7 8 9 10

BOOK DESIGN BY KAROLINA HARRIS

PREFACE

Our century inherited from the last a popular—if fruitless—debate on the relative value of science and literature in a liberal education. Nobody won! For centuries, thinking people have been fascinated by novel theories and discoveries, whether they come dressed as poems, plays, philosophical reflections, or promising experiments. Indeed, what we mean by a world view is the interaction of all these in any given age.

This collection of readings, insofar as it is a history, is a history of these interactions. It is not intended to be either a history of science or a complete chronology of literature about science. Rather, it is an encouragement to reflect on the two together and how the interaction of their language and ideas causes permutations of human thought.

Because style demonstrates most intimately the operations of thought, we have in many cases deliberately retained archaic spelling and punctuation, so that the reader might adjust to the mood, tone, and turns of expression that distinguish former ages from our own. Only recently, Stephen Jay Gould, a fine stylist and writer on evolution, commented that "science is a balanced interaction of mind and nature." His definition is equally applicable to literature, and we hope that this volume will encourage you to explore the patterns of thought that join them together.

ACKNOWLEDGMENTS

There are many people we would like to thank. First is our editor, Eunice Riedel, who gave us the initial encouragement for the book and who tended our manuscript with care and with the help of Randy Ladenheim. We would also like to thank our colleagues at C. W. Post College, Long Island University: Don Gelman, Joan Shields, and Seymour Trester for reading the astronomy, chemistry, and physics sections. We are also indebted to Sheila Martinez for doing a difficult typing job and all the reference librarians at C. W. Post for helping us forage out hidden information. To John Digby, who contributed the collage illustrations, a special thanks for illuminating the way that artistic imagination can give further life to scientific ideas.

CONTENTS

PHYSICS

CONTENTS
13

BIOLOGY

SCIENCE

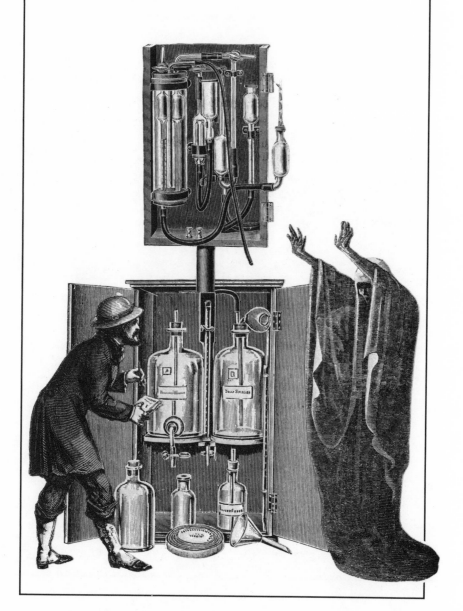

INTRODUCTION

> The discoveries of Science, the works of art are explorations—more, are explosions, of a hidden likeness.
> —Jacob Bronowski, *Science and Human Values*

When people use the word *science*, they usually mean something positive. "Let's do it scientifically" is almost equivalent to "Let's do it right." In advertisements products are presented as "scientifically designed," as if to say "rationally" and "precisely." Ironically, for this very reason science has often frightened people, people who are suspicious of dispassionate objectivity and sometimes even of progress.

In this section we present a variety of opinions about science ranging from George Herbert's tirade against probing God's unfathomable intentions to Jacob Bronowski's passionate belief that science and art discover mirror facets of imagination. Although some of these views can be understood historically, it is also possible to read them out of historical, that is, chronological, order and discover that many of the same feelings have been expressed by literary defenders and detractors of science in every generation. The ideas themselves, although subject to permutation, reflect the abiding concern of man with the operations of science.

Poets and essayists have always been willing to give their opinions on the value and proper position of science; however, in the quest for truth they have rarely discussed the more central and more difficult question of what science is. This is a question that has yet to be answered with finality by philosophers of science. While the general public may identify scientists with people wearing white lab coats—even *Sesame Street* has contributed its Dr. Bunsen Honeydew to this image—this does not help toward a precise definition of science.

For a long time it was generally assumed that science was the attempt to verify hypotheses by experimentation. This verification theory of science was the best available description for quite some time. It regarded scientists as formulating hypotheses, conducting experiments, and then retaining those hypotheses that are verified. The problem is that far too many hypotheses can be verified in this manner. For example, let us assume that every morning I get up before the sun comes up and clap my hands; soon the sun rises. I might formulate the theory that clapping my hands causes the sun to rise. As a scientist, I ask, what observation should I be able to make if my hypothesis is true? The clear answer is that if I clap my hands, the sun should subsequently rise. I repeat my little experiment of clapping my hands and indeed the sun rises. My hypothesis is confirmed; clapping my hands causes the sun to rise! By this experiment I seem to have confirmed a hypothesis that is clearly false. The reason for this curious state of affairs is that the logic is incorrect. If we diagrammed the procedure, we might say:

$H \rightarrow O$ If the hypothesis is true, then a certain observation is entailed

$\underline{\quad\quad O}$ The observation did indeed occur

$\therefore H$ Thus my hypothesis is true

Logic simply doesn't work like that. The form of the above argument is invalid. While it may be true that the hypothesis entails an observation and the observation was in fact made, this does not guarantee that my hypothesis was true. There may be reasons apart from hand clapping to explain why the sun rose.

Since there is a firmly rooted belief that science should be logical and since the validation of hypotheses by observation as presented above is invalid, this cannot be the method used in science. Sir Karl Popper, the great British philosopher of science, has presented what seems to be the best solution to the problem. His suggestion is that scientists do not attempt to confirm hypotheses, they attempt to falsify them. Hypotheses that survive the attempt at falsification are kept as possibly true. This new schema of science can be applied to our example of the correlation between the sun rising and my clapping hands. If we want to falsify this hypothesis, the real test would be to abstain from clapping hands and then observe whether the sun rises. If the sun rises in the absence of the hand clapping, then we can be certain that the clapping was not the cause of the sun's rising. This is valid reasoning. Thus, if a scientist sets up his experiments so that he can attempt to falsify a hypothesis, and the hypothesis withstands the test, then it is a candidate for the explanation of a particular phenomenon.

This falsification theory of science has many consequences that recommend it. First, it squares with our experience that science is tentative—nothing is ever finally settled, all "truths of science" are open to further testing. Also, it implies that if a hypothesis *in principle* can't be falsified, then it is not a scientific hypothesis. The hypothesis that my car is pulled by an invisible, undetectable gremlin who changes into an engine whenever anyone looks for it is not falsifiable. No matter what I observe, it will be compatible with the hypothesis. When I look for it and observe my engine, this only confirms my hypothesis. Since scientific method involves falsifying hypotheses, and this hypothesis cannot be falsified, then clearly it is not scientific. This point leads to one last conclusion, which doubting poets seem to have known for centuries but modern man has often forgotten. *Science* is not a success word.

Using scientific method—forming hypotheses, conducting experiments, making observations, etc., all in an attempt to falsify hypotheses—does not guarantee success. It is possible to be a good scientist and not make progress. The history of science is full of cases where the method was properly used but the desired results were not produced. Science is indeed a great, often exhilarating enterprise that yields results of the greatest importance to mankind, but it may not be exhaustive; there may be truths that are not amenable to scientific method. Given its great potential and equally significant limitations, it is not surprising that the literature of science is extreme in both its praise and condemnation.

In the late seventeenth century the Royal Society in England began meeting as a fellowship of scientific investigators interested in everything from meteors and mechanical air pumps to exotic birds. Abraham Cowley, a poet, physician, and founding member, eulogized the group as herculean champions of human liberty who freed the mind from the stronghold of monstrous, false beliefs. Like Francis Bacon before him, he was of the essential opinion that knowledge is power. Jonathan Swift had another, less flattering opinion, of the society, which he satirized as the Academy of Lagado in *Gulliver's Travels;* the Royal Society appeared to him as a collection of egotistical "projectors" in pursuit of ridiculous proofs irrelevant to life. While Swift regarded science as a sterile activity, it struck others, like the poet Mark Akenside, as the work of imagination. Akenside's hymn projected a romantic view of the power of science, which continued to stir many poets during the nineteenth century. Even William Wordsworth, in his famous Preface to the *Lyrical Ballads,* connected poetry's search for elemental principles with the pursuit of science, and he regarded the pleasures and discoveries of science as the poet's proper guide. Edgar Allan Poe, and in our own century Robert Frost, felt threatened by the intrusion of science on nature. To Poe its presence seemed to kill the myths; to Frost it seemed irrelevant in view of man's minute significance in the cosmos. Perhaps because of man's finite limits, as both Martin Tucker and A. R. Ammons suggest, it is difficult to face the "crux of the matter" revealed by science; yet it is there staring man in the face, and the serious intellect has always desired to come to terms with it, either by faith or reason—sometimes both.

FRANCIS BACON
(1561–1626)

From *Wisdom of the Ancients*

XXVIII. SPHINX, OR SCIENCE

Explained of the Sciences

They relate that Sphinx was a monster, variously formed, having the face and voice of a virgin, the wings of a bird, and the talons of a griffin. She resided on top of a mountain, near the city Thebes, and also beset the highways. Her manner was to lie in ambush and seize the travelers, and having them in her power, to propose to them certain dark and perplexed riddles, which it was thought she received from the Muses, and if her wretched captives could not solve and interpret these riddles, she with great cruelty fell upon them, in their hesitation and confusion, and tore them to pieces. This plague, having reigned a long time, the Thebans at length offered their kingdom to the man who could interpret her riddles, there being no other way to subdue her. Oedipus, a penetrating and prudent man, though lame in his feet, excited by so great a reward, accepted the condition, and with a good assurance of mind, cheerfully presented himself before the monster, who directly asked him, "What creature that was, which being born four-footed, afterward became two-footed, then three-footed, and lastly four-footed again?" Oedipus, with presence of mind, replied it was man, who, upon his first birth and infant state, crawled upon all four in endeavoring to walk; but not long after went upright upon his two natural feet; again, in old age walked three-footed, with a stick; and at last growing decrepit, lay four-footed confined to his bed; and having by this exact solution obtained the victory, he slew the monster, and, laying the carcass upon an ass, led her away in triumph; and upon this he was, according to the agreement, made king of Thebes.

Explanation.—This is an elegant, instructive fable, and seems invented to represent science, especially as joined with practice. For science may, without absurdity, be called a monster, being strangely gazed at and admired by the ignorant and unskillful. Her figure and form is various, by reason of the vast variety of subjects that science considers; her voice and countenance are represented female, by reason of her gay appearance and volubility of speech; wings are added, because the sciences and their inventions run and fly about in a moment, for knowledge, like light communicated from one torch to another, is presently caught and copiously diffused; sharp and hooked talons are elegantly attributed to her, because the axioms and arguments of science enter the mind, lay hold of it, fix it down, and keep it from moving or slipping away. This the sacred philosopher observed, when he said, "The words of the wise are like goads or nails driven far in."[1] Again, all science seems placed on high, as it were on the tops of mountains that are hard to climb; for science is justly imagined a sublime and lofty thing, looking down upon ignorance from an eminence, and at the same time taking an extensive view on all sides, as is usual on the tops of mountains. Science is said to beset the highways, because through all the journey and peregrination of human life there is matter and occasion offered of contemplation.

Sphinx is said to propose various difficult questions and riddles to men, which she received from the Muses; and these questions, so long as they remain with the Muses, may very well be unaccompanied with severity, for while there is no other end of contemplation and inquiry but that of knowledge alone, the understanding is not oppressed, or driven to straits and difficulties, but expatiates and ranges at large, and even receives a degree of pleasure from doubt and variety; but after the Muses have given over their riddles to Sphinx, that is, to

[1] Eccles. 12:11

practice, which urges and impels to action, choice, and determination, then it is that they become torturing, severe, and trying, and unless solved and interpreted, strangely perplex and harass the human mind, rend it every way, and perfectly tear it to pieces. All the riddles of Sphinx, therefore, have two conditions annexed, viz., dilaceration to those who do not solve them, and empire to those that do. For he who understands the thing proposed obtains his end, and every artificer rules over his work.

Sphinx has no more than two kinds of riddles, one relating to the nature of things, the other to the nature of man; and correspondent to these, the prizes of the solution are two kinds of empire—the empire over nature, and the empire over man. For the true and ultimate end of natural philosophy is dominion over natural things, natural bodies, remedies, machines, and numberless other particulars, though the schools, contented with what spontaneously offers, and swollen with their own discourses, neglect, and in a manner despise, both things and works.

But the riddle proposed to Oedipus, the solution whereof acquired him the Theban kingdom, regarded the nature of man; for he who has thoroughly looked into and examined human nature, may in a manner command his own fortune, and seems born to acquire dominion and rule. Accordingly, Virgil properly makes the arts of government to be the arts of the Romans. It was, therefore, extremely apposite in Augustus Caesar to use the image of Sphinx in his signet, whether this happened by accident or by design; for he of all men was deeply versed in politics, and through the course of his life very happily solved abundance of new riddles with regard to the nature of man; and unless he had done this with great dexterity and ready address, he would frequently have been involved in imminent danger, if not destruction.

It is with the utmost elegance added in the fable, that when Sphinx was conquered, her carcass was laid upon an ass; for there is nothing so subtile and abstruse, but after being once made plain, intelligible, and common, it may be received by the slowest capacity.

We must not omit that Sphinx was conquered by a lame

man, and impotent in his feet; for men usually make too
much haste to the solution of Sphinx's riddles; whence it happens, that she prevailing, their minds are rather racked and
torn by disputes, than invested with command by works and
effects.

ABRAHAM COWLEY

(1618–67)

To the Royal Society

1

Philosophy, the great and only Heir
    Of all that Human Knowledge, which has bin
Unforfeited by Man's rebellious Sin,
    Though full of years He do appear,
(Philosophy, I say, and call it, He,
For whatsoe'er the Painter's Fancy be,
    It a Male-Virtu seems to me)
Has still been kept in Nonage 'till of late,
Nor manag'd or enjoy'd his vast Estate:
Three or four thousand years one would have thought,
To ripeness and perfection might have brought
    A Science so well bred and nurst,
And of such hopeful parts too at the first.
But, oh, the Guardians, and the Tutors then,
(Some negligent, and some ambitious men)
    Would ne're consent to set him Free,
Or his own Natural Powers to let him see,
Lest that should put an end to their Autoritie.

2

That his own business he might quite forgit
They amus'd him with the sports of wanton Wit,

With the Desserts of Poetry they fed him,
Instead of solid meats t' encrease his force:
Instead of vigorous exercise, they led him
Into the pleasant Labyrinths of ever-fresh Discours:
    Instead of carrying him to see
The Riches which doe hoorded for him lye,
    In Nature's endless Treasurie,
    They chose his Eye to entertain
    (His curious but not covetous Eye)
With painted Scenes, and Pageants of the Brain.
Some few exalted Spirits this latter Age has shown,
That labour'd to assert the Liberty
(From Guardians, who were now Usurpers grown)
Of this Old Minor still, captiv'd Philosophy;
    But 'twas Rebellion call'd to fight
    For such a long-oppressed Right.
Bacon at last, a mighty Man, arose,
    Whom a wise King and Nature chose,
    Lord Chancellour of both their Laws,
And boldly undertook the injur'd Pupil's caus.

3

Autority, which did a Body boast,
Though 'twas but Air condens'd and stalk'd about,
Like some old Giant's more Gigantic Ghost,
    To terrifie the learned Rout
With the plain Magique of true Reason's Light,
    He chac'd out of our sight;
Nor suffer'd living Men to be misled
    By the vain shadows of the Dead:
To Graves, from whence it rose, the conquer'd Phantome fled.
    He broke that Monstrous God which stood
In midst of th' Orchard, and the whole did claim,
    Which with a useless Sith of Wood,
    And something else not worth a name,
    (Both vast for shew, yet neither fit
    Or to Defend, or to Beget;
    Ridiculous and senceless Terrors!) made
Children and superstitious Men afraid.
    The Orchard's open now, and free;

Bacon has broke that Scar-crow Deitie;
   Come, enter, all that will,
Behold the rip'ned Fruit, come gather now your Fill.
   Yet still, methinks, we fain would be
   Catching at the Forbidden Tree,
   We would be like the Deitie,
When Truth and Falshood, Good and Evil, we
Without the Sences aid within our selves would see;
   For 'tis God only who can find
   All Nature in his Mind.

4

From Words, which are but Pictures of the Thought,
(Though we our Thoughts from them perversly drew)
To Things, the Mind's right Object, he it brought;
Like foolish Birds to painted Grapes we flew;
He sought and gather'd for our use the Tru;
And when on heaps the chosen Bunches lay,
He prest them wisely the Mechanic way,
'Till all their juyce did in one Vessel joyn,
Ferment into a Nourishment Divine,
   The thirsty Soul's refreshing Wine.
Who to the life an exact Piece would make,
Must not from other's Work a Copy take;
   No, not from Rubens or Vandike;
Much less content himself to make it like
Th' Idaeas and the Images which ly
In his own Fancy, or his Memory.
   No, he before his Sight must place
   The Natural and Living Face;
   The real Object must command
Each judgment of his Eye, and Motion of his Hand.

5

From these and all long Errors of the Way,
In which our wandring Predecessors went.
And like th' old Hebrews many Years did stray,
   In Desarts but of small extent,

Bacon, like Moses, led us forth at last;
  The barren Wilderness he past,
  Did on the very Border stand
  Of the blest promis'd Land,
And from the Mountains Top of his Exalted Wit,
  Saw it himself, and shew'd us it.
But Life did never to one Man allow
Time to Discover Worlds, and Conquer too;
Nor can so short a Line sufficient be
To fadome the vast depths of Nature's Sea:
  The work he did we ought t' admire,
And were unjust if we should more require
From his few years, divided 'twixt th' Excess
Of low Affliction, and high Happiness
For who on things remote can fix his Sight,
That's always in a Triumph or a Fight?

6

From you, great Champions, we expect to get
These spacious Countries but discover'd yet;
Countries where yet instead of Nature, we
Her images and Idols worship'd see:
These large and wealthy Regions to subdu,
Though Learning has whole Armies at command.
  Quarter'd about in every Land,
A better Troop she ne're together drew.
  Methinks, like Gideon's little Band,
  God with Design has pickt out you,
To do these noble Wonders by a Few:
When the whole Host he saw, They are (said he)
  Too many to Orecome for Me;
  And now he chuses out his Men,
  Much in the way that he did then:
  Not those many, whom he found
  Idlely extended on the ground,
  To drink with their dejected head,
The Stream, just so as by their Mouths it fled:
No, but those Few who took the Waters up,
And made of their laborious Hands the Cup.

7

Thus you prepar'd; and in the glorious Fight
    Their wondrous pattern too you take:
Their old and empty Pitchers first they brake,
And with their Hands then lifted up the Light.
    Io! Sound too the Trumpets here!
Already your victorious Lights appear;
New Scenes of Heaven already to espy,
And Crowds of golden Worlds on high;
Which from the spacious Plains of Earth and Sea,
    Could never yet discover'd be,
By Sailors or Chaldaeans watchful Eye.
Nature's great Works no Distance can obscure,
No smalness her near Objects can secure;
    Y'have taught the curious Sight, to press
    Into the privatest recess
Of her imperceptible Littleness.
She with much stranger Art than his who put
    All th' Iliads in a Nut,
The numerous works of life does into atomes shut.
    Y' have learn'd to Read her smallest Hand,
And well begun her deepest Sense to Understand.

8

Mischief and tru Dishonour fall on those,
Who would to laughter or to scorn expose
So Virtuous and so Noble a Design,
So human for its Use, for Knowledge so Divine.
The things which these proud men despise, and call
    Impertinent, and vain, and small,
Those smallest things of Nature let me know
Rather than all their greatest Actions Doe,
Whoever would Deposed Truth advance
    Into the Throne usurp'd from it,
Must feel at first the Blows of Ignorance,
    And the sharp Points of Envious Wit.
So when, by various turns of the Celestial Dance,

In many thousand years,
　A Star, so long unknown, appears,
Tho' Heaven it self more beauteous by it grow,
It troubles and alarms the World below,
Does to the Wise a Star, to Fools a Meteor show.

9

With Courage and Success you the bold work begin;
　Your Cradle has not Idle bin:

None e're but Hercules and you could be
At five years Age worthy a History.
　And ne're did Fortune better yet
　Th' Historian to the Story fit:
　As you from all Old Errors free
And purge the Body of Philosophy;
　So from all Modern Follies He
Has vindicated Eloquence and Wit.
His candid Stile like a clean Stream does slide,
　And his bright Fancy all the way
　Does, like the Sun-shine in it play;
It does like Thames, the best of Rivers, glide,
Where the God does not rudely overturn,
　But gently pour the Crystal Urn,
And with judicious hand does the Whole Current guide.
'T has all the Beauties Nature can impart,
And all the comely Dress, without the paint of Art.

GEORGE HERBERT

(1593–1633)

Vanitie

The fleet Astronomer can bore,
And thred the spheres with his quick-piercing minde:
He views their stations, walks from doore to doore,
   Surveys, as if he had design'd
To make a purchase there: he sees their dances,
   And knoweth long before
Both their full-ey'd aspects, and secret glances.

The nimble Diver with his side
Cuts through the working waves, that he may fetch
His dearly-earned pearl; which God did hide
   On purpose from the ventrous wretch;
That he might save his life, and also hers,
   Who with excessive pride
Her own destruction and his danger wears.

The subtil Chymick can devest
And strip the creature naked, till he finde
The callow principles within their nest:
   There he imparts to them his minde,
Admitted to their bed-chamber, before
   They appeare trim and drest
To ordinarie suitours at the doore.

What hath not man sought out and found,
But his deare God? Who yet his glorious law

Embosomes in us, mellowing the ground
　With showres and frosts, with love & aw,
So that we need not say, Where's this command?
　Poore man, thou searchest round
To find out *death*, but missest *life* at hand.

ALEXANDER POPE
(1688–1744)

From *An Essay on Man*

BOOK II

　Go, wondrous creature! mount where Science guides,
Go, measure earth, weigh air, and state the tides;
Instruct the planets in what orbs to run,
Correct old Time, and regulate the Sun;
Go, soar with Plato to th' empyreal sphere,
To the first good, first perfect, and first fair;
Or tread the mazy round his followers trod,
And quitting sense call imitating God;
As Eastern priests in giddy circles run,
And turn their heads to imitate the Sun.
Go, teach Eternal Wisdom how to rule—
Then drop into thyself, and be a fool!

JONATHAN SWIFT
(1667–1745)

From *Gulliver's Travels*

BOOK III, Chapter 5
THE ACADEMY OF SCIENCE AT LAGADO

The first man I saw was of a meagre aspect, with sooty hands and face, his hair and beard long, ragged and singed in several places. His clothes, shirt, and skin were all of the same colour. He had been eight years upon a project for extracting sun-beams out of cucumbers, which were to be put into vials hermetically sealed, and let out to warm the air in raw inclement summers. He told me, he did not doubt in eight years more he should be able to supply the Governor's garden with sunshine at a reasonable rate; but he complained that his stock was low, and entreated me to give him something as an encouragement to ingenuity, especially since this had been a very dear season for cucumbers. I made him a small present, for my Lord had furnished me with money on purpose, because he knew their practice of begging from all who go to see them.

I went into another chamber, but was ready to hasten back, being overcome with a horrible stink. My conductor pressed me forward, conjuring me in a whisper to give no offence, which would be highly resented, and therefore I durst not so much as stop my nose. The projector of this cell was the most ancient student of the Academy. His face and beard were of a pale yellow; his hands and clothes daubed over with filth. When I was presented to him, he gave me a very close embrace (a compliment I could well have excused).

His employment from his first coming into the Academy was an operation to reduce human excrement to its original food, by separating the several parts, removing the tincture which it receives from the gall, making the odour exhale, and scumming off the saliva. He had a weekly allowance from the society of a vessel filled with human ordure, about the bigness of a Bristol barrel.

I saw another at work to calcine ice into gunpowder, who likewise showed me a treatise he had written concerning the malleability of fire, which he intended to publish.

There was a most ingenious architect who had contrived a new method for building houses, by beginning at the roof and working downwards to the foundation, which he justified to me by the like practice of those two prudent insects, the bee and the spider.

*　　*　　*

There was an astronomer who had undertaken to place a sundial upon the great weathercock on the town-house, by adjusting the annual and diurnal motions of the earth and sun, so as to answer and coincide with all accidental turnings by the wind.

I was complaining of a small fit of colic, upon which my conductor led me into a room, where a great physician resided, who was famous for curing that disease by contrary operations of the same instrument. He had a large pair of bellows with a long slender muzzle of ivory. This he conveyed eight inches up the anus, and drawing in the wind, he affirmed he could make the gust as lank as a dried bladder. But when the disease was more stubborn and violent, he let in the muzzle while the bellows were full of wind, which he discharged into the body of the patient, then withdrew the instrument to replenish it, clapping his thumb strongly against the orifice of the fundament; and this being repeated three or four times, the adventitious wind would rush out, bringing the noxious along with it (like the water put into a pump) and the patient recovers. I saw him try both experiments upon a dog. . . . The dog died on the spot, and we left the doctor endeavouring to recover him by the same operation.

MARK AKENSIDE

(1721–70)

Hymn to Science

1   Science! thou fair effusive ray
From the great source of mental day,

    Free, generous, and refined!
Descend with all thy treasures fraught,
Illumine each bewilder'd thought,
    And bless my labouring mind.

2   But first with thy resistless light,
Disperse those phantoms from my sight,
    Those mimic shades of thee:
The scholiast's learning, sophist's cant,
The visionary bigot's rant,
    The monk's philosophy.

3   Oh! let thy powerful charms impart
The patient head, the candid heart,
    Devoted to thy sway;
Which no weak passions e'er mislead,
Which still with dauntless steps proceed
    Where reason points the way.

4   Give me to learn each secret cause;
Let Number's, Figure's, Motion's laws
    Reveal'd before me stand;
These to great Nature's scenes apply,
And round the globe, and through the sky,
    Disclose her working hand.

5   Next, to thy nobler search resign'd,
The busy, restless, Human Mind
    Through every maze pursue;
Detect Perception where it lies,
Catch the Ideas as they rise,
    And all their changes view.

6   Say from what simple springs began
The vast ambitious thoughts of man,
    Which range beyond control,
Which seek eternity to trace,
Dive through the infinity of space,
    And strain to grasp the whole.

7   Her secret stores let Memory tell,
Bid Fancy quit her fairy cell,
    In all her colours dress'd;
While prompt her sallies to control,
Reason, the judge, recalls the soul
    To Truth's severest test.

8   Then launch through Being's wide extent;
Let the fair scale with just ascent
    And cautious steps be trod;
And from the dead, corporeal mass,
Through each progressive order pass
    To Instinct, Reason, God.

9   There, Science! veil thy daring eye;
Nor dive too deep, nor soar too high,
    In that divine abyss;
To Faith content thy beams to lend,
Her hopes to assure, her steps befriend
    And light her way to bliss.

10   Then downwards take thy flight again,
Mix with the policies of men,
    And social Nature's ties;
The plan, and genius of each state,
Its interest and its powers relate,
    Its fortunes and its rise.

11   Through private life pursue thy course,
Trace every action to its source,
   And means and motives weigh:
Put tempers, passions, in the scale;
Mark what degrees in each prevail,
   And fix the doubtful sway.

12   That last best effort of thy skill,
To form the life, and rule the will,
   Propitious power! impart:
Teach me to cool my passion's fires,
Make me the judge of my desires,
   The master of my heart.

13   Raise me above the vulgar's breath,
Pursuit of fortune, fear of death,
   And all in life that's mean:
Still true to reason be my plan,
Still let my actions speak the man,
   Through every various scene.

14   Hail! queen of manners, light of truth;
Hail! charm of age, and guide of youth;
   Sweet refuge of distress:
In business, thou! exact, polite;
Thou giv'st retirement its delight,
   Prosperity its grace.

15   Of wealth, power, freedom, thou the cause;
Foundress of order, cities, laws,
   Of arts inventress thou!
Without thee, what were human-kind?
How vast their wants, their thoughts how blind!
   Their joys how mean, how few!

16   Sun of the soul! thy beams unveil:
Let others spread the daring sail

On Fortune's faithless sea:
While, undeluded, happier I
From the vain tumult timely fly,
   And sit in peace with thee.

WILLIAM WORDSWORTH

(1770–1850)

From Preface to *Lyrical Ballads*

Poetry is the image of man and nature. The obstacles which stand in the way of the fidelity of the biographer and historian, and of their consequent utility, are incalculably greater than those which are to be encountered by the poet who comprehends the dignity of his art. The poet writes under one restriction only, namely, the necessity of giving immediate pleasure to a human being possessed of that information which may be expected from him, not as a lawyer, a physician, a mariner, an astronomer, or a natural philosopher, but as a man. Except this one restriction, there is no object standing between the poet and the image of things; between this, and the biographer and historian, there are a thousand.

Nor let this necessity of producing immediate pleasure be considered as a degradation of the poet's art. It is far otherwise. It is an acknowledgment of the beauty of the universe, an acknowledgment the more sincere because not formal, but indirect; it is a task light and easy to him who looks at the world in the spirit of love: further, it is a homage paid to the native and naked dignity of man, to the grand elementary principle of pleasure, by which he knows, and feels, and lives, and moves. We have no sympathy but what is propagated by pleasure: I would not be misunderstood; but wherever we sympathize with pain, it will be found that the sympathy is produced and carried on by subtle combinations with pleasure. We have no knowledge, that is, no general principles drawn from the contemplation of particular facts, but what has been built up by pleasure, and exists in us by pleasure alone. The man of science, the chemist and mathematician, whatever difficulties and disgusts they may have had to struggle with, know and feel this. However painful may be the objects with which the anatomist's knowledge is connected, he feels that his knowledge is pleasure; and where he has no pleasure he has no knowledge. What then does the poet? He considers man and the objects that surround him as acting and reacting upon each other, so as to produce an infinite complexity of pain and pleasure; he considers man in his own nature and in his own ordinary life as contemplating this with a certain quantity of immediate knowledge, with certain convictions, intuitions, and deductions, which from habit acquire the quality of intuitions; he considers him as looking upon this complex scene of ideas and sensations, and finding everywhere objects that immediately excite in him sympathies which, from the necessities of his nature, are accompanied by an overbalance of enjoyment.

To this knowledge which all men carry about with them, and to these sympathies in which, without any other discipline than that of our daily life, we are fitted to take delight, the poet principally directs his attention. He considers man and nature as essentially adapted to each other, and the mind of man as naturally the mirror of the fairest and most interesting properties of nature. And thus the poet, prompted by this feeling of pleasure, which accompanies him through the whole course of his studies, converses with general nature, with affections akin to those which, through labor and length of time, the man of science has raised up in himself, by conversing with those particular parts of nature which are the objects of his studies. The knowledge both of the poet and the man of science is pleasure; but the knowledge of the one cleaves to us as a necessary part of our existence, our natural and unalienable inheritance; the other is a personal and individual acquisition, slow to come to us, and by no habitual and direct sympathy connecting us with our fellow-beings. The man of science seeks truth as a remote and unknown benefactor; he cherishes and loves it in his solitude: the poet, singing a song in which all human beings join with him, rejoices in the presence of truth as our visible friend and hourly companion. Poetry is the breath and finer spirit of all knowledge; it is the impassioned expression which is in the countenance of all science. Emphatically may it be said of the poet, as Shakespeare hath said of man, "that he looks before and after."[2] He is the rock of defense for human nature; an upholder and preserver, carrying everywhere with him relationship and love. In spite of difference of soil and climate, of language and manners, of laws and customs: in spite of things silently gone out of mind, and things violently destroyed; the poet binds together by passion and knowledge the vast empire of human society, as it is spread over the whole earth, and over all time. The objects of the poet's thoughts are everywhere; though the eyes and senses of man are, it is true, his favorite guides, yet he will follow wheresoever he can find an atmosphere of sensation in which to move his wings. Poetry is the first and last of all knowledge—it is as immortal as the heart of man. If the labors of men of science should ever create any material revolution, direct or indirect, in our condition, and in the impressions which we habitually receive, the poet will sleep then no more than at present; he will be ready to follow the steps of the man of science, not only in those general indirect effects, but he will be at his side, carrying sensation into the midst of the objects of the science itself. The remotest discoveries of the chemist, the botanist, or mineralogist, will be as proper objects of the poet's art as any upon which it can be employed, if the time should ever come when these things shall be familiar to us, and the relations under which they are contemplated by the followers of these respective sciences shall be manifestly and palpably material to us as enjoying and suffering beings. If the time should ever come when what is now called science, thus familiarized to men, shall be ready to put on, as it were, a form of flesh and blood, the poet will lend his divine spirit to aid the transfiguration, and will welcome the being thus produced as a dear and genuine inmate of the household of man. It is not, then, to be supposed that any one who holds that sublime notion of poetry which I have attempted to convey, will break in upon the sanctity and truth

[2]*Hamlet*, iv. 4. 37.

of his pictures by transitory and accidental ornaments, and endeavor to excite admiration of himself by arts, the necessity of which must manifestly depend upon the assumed meanness of his subject.

EDGAR ALLAN POE

(1809–49)

Sonnet—To Science

Science! true daughter of Old Time thou art!
    Who alterest all things with thy peering eyes.
Why preyest thou thus upon the poet's heart,
    Vulture, whose wings are dull realities?
How should he love thee? or how deem thee wise?
    Who wouldst not leave him in his wandering
To seek for treasure in the jewelled skies,
    Albeit he soared with an undaunted wing?
Hast thou not dragged Diana[3] from her car?
    And driven the Hamadryad[4] from her flood,
The Elfin from the green grass, and from me
The summer dream beneath the tamarind tree?

[3]Moon.
[4]Water spirit.

ROBERT FROST
(1875–1963)

Why Wait for Science

Sarcastic Science she would like to know,
In her complacent ministry of fear,
How we propose to get away from here
When she has made things so we have to go
Or be wiped out. Will she be asked to show
Us how by rocket we may hope to steer
To some star off there say a half light-year
Through temperature of absolute zeró?
Why wait for science to supply the how
When any amateur can tell it now?
The way to go away should be the same
As fifty million years ago we came—
If anyone remembers how that was.
I have a theory, but it hardly does.

MARTIN TUCKER
(1928—)

Poetry and the New Science

"I didn't get an atom of what he said,"
I heard him say the other day,
and I fell to wondering how precise

common speech can teach us
wonders of the spirit.

My friend rarely examines facts
for what they seem.
His interest lies in surface things:
nine cats lying on a microscope,
or knocking on Dr. Johnson's favorite beam.[5]

Yet with science dabbling in the foreseen,
and facts weaned with creative imagining,
he finds his life gone to unsettling.
He yearns for categories that have walls,
and labels he can affix without
                              divining anything.

A. R. AMMONS
(1926—)

Exotic

Science outstrips
other modes &
reveals more of
the crux of the matter
than we can calmly
handle

---

[5]Dr. Samuel Johnson (1709–84), who refuted Bishop Berkeley's Idealism by kicking a
solid beam.

ASTRONOMY

INTRODUCTION

Therefore let us permit these new hypotheses to make a public appearance among old ones which are themselves no more probable, especially since they are wonderful and easy and bring with them a vast storehouse of learned observations. And as far as hypotheses go, let no one expect anything in the way of certainty from astronomy, since astronomy can offer us nothing certain, lest, if anyone take as true that which has been constructed for another use, he go away from this discipline a bigger fool than when he came to it. Farewell.

—Copernicus

For many historians of science, science begins with Alexander the Great. After Alexander conquered the oriental powers of the ancient world there was a period of stability that produced sciences unlike those developed previously. Foremost among these Hellenistic sciences was astronomy, the science of heavenly bodies. While it is traditional to grant that any period of science owes a great debt to earlier periods, this is not so with Hellenistic astronomy. The astronomy that developed in Alexandria was so far ahead of what the Egyptians and Babylonians had that it would be futile to try to trace Greek astronomy to such sources.

In the case of the Egyptians there is only one significant astronomical contribution creditable to them: the calendar. At first the Egyptians used a calendar consisting of 12 months of 30 days, giving a 360-day year. Soon the Egyptians realized that this calendar did not square with what occurred in nature. Indeed, one of their three seasons was called "inundation," because the Nile overflowed its banks, but since each year of the calendar was approximately five and a quarter days off, eventually the season of inundation came when the Nile was at its lowest point! To correct for such incongruities the Egyptians added 5 days to the end of the year to give them a 365-day calendar, which the Greeks adopted. This, the only Egyptian contribution to Hellenic astronomy, was not made on the basis of careful scientific observation and recording. Rather, any observer could have seen that the first calendar did not accurately describe nature, and from such an observation it was a simple matter to record for each year how far out of accord with nature it was and then with arithmetic determine what correction was necessary. Thus the Egyptian contribution was quite far from what one might call "exact science."

The Egyptians were, however, stargazers, and did record the heavens as they saw them, but without any scientific method in mind. Stars were depicted in many of the tombs of the nobles in Luxor, and in the Middle Kingdom star representations were popular inside sarcophagus lids. But this too was far from science.

Babylonian astronomy, contrary to popular belief, was in some ways even less sophisticated than the Egyptians'. Their astronomy was not empirically based or oriented, but was more theoretical and mathematical. The astronomical archives both at Uruk and Babylon have yielded thousands of astronomical tablets, but they are not really the precursors of the kind of astronomy the Greeks were to develop. The Babylonians put all celestial phenomena on an equal footing. Thus clouds, which are a meteorological phenomenon, were considered similar to planets, stars, etc., which are astronomical phenomena. A constant concern in Babylonian astronomy was discovering an omen in the sky that could enable one to predict the future.

In the Greek astronomy prior to Alexander there was a vestige of this treatment of all heavenly phenomena as similar. The early Greek astronomers tried to explain all things in the sky by analogy to the behavior of familiar objects on earth. Such attempts at explanation seem to have arisen around 600 B.C. One of the earliest Greeks concerned with the heavens was Thales, who supposedly traveled to Egypt, where he learned a great deal. It is difficult to imagine what he could have learned there, and unfortunately his writings have been lost, so that all we have is the tradition and not the evidence.

Anaximander was the first Greek whose astronomical views were preserved (by Theophrastos, a pupil of Aristotle). He presents what is perhaps the first description of gravitational attraction. He says that the earth swings freely in space "because of its equal distance from everything else." He hit upon one of the two components of gravitational attraction—distance. He did not realize that even if the earth were equidistant from everything else, it would be necessary for the masses of the things exerting opposite pulls to be equal. Otherwise the earth would be pulled in the direction of the greatest mass.

Anaximander believed that the earth was a cylinder and that man lived on the top flat surface. About one hundred years later Pythagoras (c. 580–500 B.C.), while giving an accurate account of the the causes of eclipses, described the earth as a sphere. Further, he claimed that the earth was not in the center of the universe; it was fire that occupied this place. He probably did not mean the sun, but rather that fire was a worthier element than earth and thus deserving of a more central place. The selection from Aristotle presents the Pythagorean position and points out that on the issue of centrality the Pythagoreans were not arguing from observation.

The concept of astronomy as a theoretical endeavor, not essentially rooted in firsthand observation, was common in Greece prior to Alexander. In Plato's *Republic,* when Socrates discusses how the youth of the city ought to be educated, he concedes that arithmetic and geometry are essential. Then, when considering astronomy as a third candidate for study, he voices the opinion that astronomy is important but the heavens should be left alone! It is only in the abstract that true eternal knowledge can be obtained, he argues. Socrates presents the belief that the laws governing the motions of the heavenly bodies can be derived without observation. One need know only certain fundamental principles and the rest will follow.

Plato (428–348 B.C.), Socrates' pupil, seems to have closely followed his teacher's views. He too was an abstractionist, at times mystical, having little respect for empirical data. Such a position allows for unfounded speculation and often leads to fantastic descriptions of the universe, such as his allegory in the *Timaeus* conceptualizing the universe as a created living being. Later, in the *Republic,* Plato relates the myth of Er, a soldier who died on the battlefield and who, with other spirits, was taken on an otherworldly journey. Here he describes the universe as revolving around a Spindle of Necessity. Attached to the spindle are eight concentric globes. Each of these worlds has fixed to it some astronomical entity. The outermost is spangled with stars; the next has the sun, the next the moon, then Saturn, Mercury, etc. Plato talks about the sounds that these bodies make as they go around in their orbits. Each, he argues, has a single tone, but together they produce a harmony. This notion of the harmony of the spheres would be criticized by Plato's successor, Aristotle (384–322 B.C.), but would be revived fifteen hundred years later by Johannes Kepler, who would be preoccupied by this concept for most of his career.

Aristotle was critical of Plato's views primarily because Plato was not an empiricist. It did not seem as if his theories matched up with experience. While Plato had written over the doorway of his academy that no man who had not studied geometry should enter, Aristotle would have preferred that students simply open their eyes. Observation was essential to his astronomical deductions. In his treatise *On the Heavens,* he says that not only the stars but the entire heavens seem to move and thus cannot possibly be at rest, as some had claimed. Thus, either both move or only one moves. His conclusion is that fixed stars are attached to circles that move.

Aristotle rejects the notion of celestial harmony on empirical grounds. He says that it does not accord with experience for we don't hear the supposed harmony. The reason that the stars do not make any sound is that they are fixed to their circles, and objects fixed to something do not rattle and make noise! According to Aristotle's theory, anything moving through air at a great speed should make a noise and be heard. He never explains why the circles, which do move through air, don't make noise. He seems, however, to be unaware of this difficulty.

He was, though, aware of the difference between astronomy and meteorology and was the first to separate the two. He wrote a distinct work, *Meteorology*, to explain natural phenomena in the heavens. Like his predecessors, Aristotle believed that the universe was made up of fire, air, earth, and water, and his meteorology was based on this assumption.

During Aristotle's time Athens had been the center of learning for the Greek city-states. After his death this changed steadily. Scholarship spread out across Greece and became the province of isolated individuals. This remained true until Alexander the Great conquered the area and at his name-city established the famous museum and library of Alexandria. The library is reputed to have contained half a million volumes, and it made Alexandria the focal point of all science.

Four hundred years after the death of Alexander, Claudius Ptolemy did research in the library. Ptolemy was born around A.D. 85 and died at the age of eighty. His major work was the *Mathematical Concordance of Astronomy*, which is better known as *Almagest*, a Latin-Greek-Arabic corruption of "greatest." In the *Almagest* Ptolemy left behind Aristotle's concern for physics and instead was primarily concerned with the *appearance* of the heavens, not with the *causes* of the motions observed. Ptolemy's description took the form of geometrical constructions. The planets moved in circles, but Ptolemy had observed that at times it seemed as if the planets went backward; circles alone would not describe these "retrograde" motions as observed. For this reason Ptolemy invoked the notion of epicycles. According to this structure, the planets move in great circles, but while they are doing this, they also move more quickly in smaller ones. It is interesting that Ptolemy did not feel constrained to find one description that would hold for every planet. Instead, he derived for each a new model, using different combinations of greater and lesser circles. The notion of a single explanatory principle governing the motion of all heavenly bodies would have to wait for the genius of Sir Isaac Newton.

The *Almagest* was Ptolemy's major astronomical work, but

a second work, *Tetrabiblos (Four Books)*, was his major astrological work. Ptolemy believed that only a fool would deny that the heavens had an effect on what occurs on earth. He observes that the moon has an effect on the tides, arguing that if accurate data could be obtained on the position of the heavens, one could begin to see even greater correlations. Among these correlations he discusses the birth of twins and deformed children, both of which he attributes to the heavens. For Ptolemy astronomy and astrology are related sciences. This liaison, developed out of general interest in heavenly causes, was to continue for well over a thousand years. Only when Newton showed the underlying physical connections between the heavens and earth would the two disciplines become separate.

For almost fourteen hundred years there were few advances in astronomy over the system proposed by Ptolemy. The Greeks turned their attention away from astronomy, and during the eighth century the Arab world became the custodians of scientific knowledge, although the Arabs did not appreciably alter the Greek tradition. Eventually, Arabic manuscripts made their way into Europe, but these manuscripts were often thirdhand. By the time they were translated into Latin they were sometimes hopelessly garbled. Thus, for centuries the major European effort was to restore Greek knowledge, not improve on it.

Nicolaus Copernicus (1473–1543) found the Ptolemaic system unacceptable, not because of any observation that contradicted it but because it was intellectually unsatisfactory. It seemed unreasonable to him that the planets should each move to a different drummer and not behave in an organized, systematic manner. For this reason he was determined to replace Ptolemy's haphazard construction.

Copernicus' system required that the sun and not the earth be regarded as the center of the universe. The theological ramifications of this are well known. Man, who was created in God's image, was no longer the focal point; divine order was disturbed. Copernicus' concern with these issues is evident in his introduction to *De revolutionibus orbium coelestium (The Revolution of the Heavenly Spheres)*, which was dedicated to Pope Paul III.

The Revolution of the Heavenly Spheres swiftly presents the entire Copernican system. One important consequence of placing the sun in the center of the universe is that it required the earth to move. In Copernicus' time there was a tradition of forceful arguments demonstrating that the earth could not be moving—there would be great winds, etc. In the first book of *De revolutionibus* Copernicus presents his system, the arguments against it, and then his replies to the critics. The work appeared in 1543, the year of Copernicus' death. On his death bed Copernicus was handed a copy of the new book but was unable to read it because of failing eyesight. Others read it with fervor, however. After thirteen hundred years astronomy, like the earth, was stirred to movement. Its next great champion was to be Johannes Kepler (1571–1630), whose discoveries required the preliminary work of Tycho Brahe (1546–1601).

Brahe was a wealthy Danish aristocrat who had sufficient funds to establish the best-equipped laboratory in the world. From his observatory on his own island he made the most precise observations of the heavens up to his time. Indeed, precise and *prolonged* observation was Tycho's contribution to astronomy. His observatory, however, was quite eccentric. Brahe himself was a strange figure. He had an artificial gold and silver nose, his own having been cut off in a duel with a fellow student over who was the better mathematician; in the basement of his observatory he kept his own private jail for residents of his island who fell out of favor; he also kept a pet dwarf named Jepp, who stayed under the table and was fed on scraps.

Despite this insanity, Brahe's work was important. As a result of his careful observations, three fundamental beliefs about the heavens were challenged: The first discrepency he discovered was the absence of stellar parallax. The patterns of the constellations remain unchanged even though the earth travels over a hundred million miles through the sky. This motion should produce a change in the way the constellations appear—unless the stars are vast distances away from the earth. Rather than revise the idea of how distant the stars are, Brahe concluded that the earth does not move. His system was thus a cross between Ptolemy and Copernicus. He placed the sun in the center but made five planets and the sun move around the earth.

The second challenge came in 1572 in the form of a new, bright star that appeared. Tycho's observation of it argued against the idea that the sphere of fixed stars was immutable.

The comet of 1577 challenged yet another traditional belief about the heavens—that the spheres on which the planets revolved were solid. It appeared that the comet had traveled through this region; therefore, the region had to be penetrable. The observations of Brahe pointed out discrepancies in Copernicus' theory and opened up the system to speculation.

The intense astronomical activity of the sixteenth century quite rapidly filtered down to the literary world. Some of the most amusing literary speculations came from Robert Burton and were published in his *Anatomy of Melancholy* (1621). Burton, a clergyman, made himself dizzy with heavenly calculations not only on the possibility of other "planetary inhabitants" but the "plurality of worlds." His was clearly a mind receptive to the mystical component of astronomy, although he shows how an overdose of abstract musing could lead a man to melancholy. Shakespeare was suspicious of "mystical astronomy," which he employs as a metaphor for the love argument in "Sonnet 14." His "astronomy"—the prediction of fortunes and natural phenomena—is what we would call astrology. Shakespeare himself brushed it off as a false knowledge compared with that derived from his lover's eyes. These, he argues, are "constant stars" by contrast with the heavens; and these are the stars by which he chooses to prognosticate.

Like Burton and Shakespeare, Kepler drew thin lines between astronomy and astrology, fact and fiction. He was, on the one hand, an assistant and disciple of Tycho Brahe and, on the other, a court astrologer and the son of a woman accused of witchcraft. These polarities exerted strong positive counterforces on his vibrant imagination. Thinly veiled in autobiographical references, his *Somnium*, or *Lunar Astronomy*, was one of the first fictional voyages to the moon. Essentially a dream fantasy of direct and sensuous contact with the universe, it epitomizes Kepler's own need not merely to calculate and know but to experience directly the power and harmony of cosmic structure.

Kepler's hunger for explanation persisted through his work. When Tycho Brahe was an old man, Kepler was just beginning to show his promise. He realized that to improve on the Copernican system he needed observations, and Tycho had them. He hired himself out as Tycho's assistant, which, however, failed to get him the essential data. Tycho was a miser with respect to his treasured observations and refused to give Kepler access to them. Rather, he doled out the information bit by bit. However, when Tycho put Kepler to work on the problem of Mars' perturbations, he was forced to hand over the Mars data. This relationship between Tycho and Kepler lasted about eighteen months, for Brahe died in 1601, and eventually Kepler became his successor and heir to the data.

The problem of describing the orbit of Mars led to one of the three laws of astronomy for which Kepler is famous. When he took observed positions of the planet, he found that it was difficult to get a fix that accommodated all the observations. He worked on this problem for eight years and published the results in his *New Astronomy* (1609). In it he declared that Mars circumscribed an oval path. This meant that the planets did not move in perfect circles or epicycles within circles, but in egg-shaped orbits. The discovery of this fact is a great tribute to the integrity of Kepler. He never ignored any of the data—Tycho's prized observations. It would have been a simple matter for him to add a few new epicycles to form a circular orbit for Mars and be done with it. But this would not have been honest.

Kepler's first law, that the planets traveled in elliptical orbits, was actually the second of the three laws that he formulated. Kepler's second law was chronologically his first and was derived at a time when he still believed that the planets traveled in circular orbits. Kepler sought to formulate a law that would describe the forces that were responsible for their movements. Physics, the science of force, matter, motion, and energy, paved the way for this new conceptual thinking in astronomy. Kepler was directing his thought to the creation of a celestial physics; he wanted to know what forces were responsible for the planetary phenomena that he observed. From observation, for example, Kepler knew that the distant planets moved more slowly than the ones closer to the sun.

He concluded that some force from the sun, a force that diminishes over distance, propels the planets. This theory for the first time proposes a continuously acting physical force rather than a clocklike universe set in motion and simply permitted to run.

Kepler's second law, derived from this concept of a continuously acting force, describes the speed of the planets. Kepler concluded that the area swept over by a line connecting a planet and the sun was a measure of the time required by the planet to traverse a particular arc. Thus, the time it takes a planet to go from A_1 to B_1 is the same as it takes to go from A to B because areas ABS and $A_1B_1S_1$ are equal.

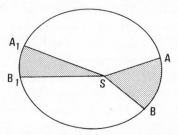

This allows for an equation describing and explaining why the planets move at different speeds in their orbits. This equation was later revised in accordance with Kepler's discovery of elliptical orbits.

The third law worked out a relationship between the period of a planet (the time it takes for one complete revolution around the sun) and its distance from the sun.

$$(\text{planetary year})^2 \propto (\text{distance from the sun})^3$$

Saturn, for example, takes thirty years to go around the sun, and Jupiter takes twelve years. By squaring the cube root of these periods one can determine the proper proportions of their distances from the sun.

Kepler is remembered for these three laws, which are still regarded as valid and "correct." But even his correct conclusions are riddled with the uncanny. In working out the orbit of Mars he made three major arithmetic mistakes, which happened to cancel one another out and yield the correct result. Throughout Kepler's work there is an element of mysticism. Frequently, his "mistakes" expose the most interesting elements of his intuitive imagination.

His mystical bent came out early in his work when in 1596 he published *Mysterium Cosmographicum (Cosmic Mystery)*. It was a Pythagorean attempt to describe a Copernican solar system. He was looking for mathematical (geometrical) laws to explain the distances between the planets. Since Plato's time it had been believed that there were five regular solids: octahedron, icosahedron, dodecahedron, tetrahedron, and cube. Kepler believed that the distance between the planets could be described in such terms. The universe in this vision became like a Chinese box with regular solids nestled one inside the other. Outside each solid revolved a planet.

Yet even this was not Kepler's most mysterious conceptualizing about the solar system. In 1619 he published the *Harmonice Mundi (Harmonies of the World)* in which he suggests a musical harmony of the cosmos. Here he reasons that the planets make sounds as they travel in their orbits and that together these sounds produce a harmony. The key to their relationship, he postulates, is the angular velocities of the planets, which forms a harmonic sequence.

Poets less concerned with mathematical diagrams were nevertheless inspired by this metaphor of perfection and the "music of the spheres" became a commonplace among their regular stock of images, as we can see in Robert Herrick's charming lyric. A long poetic expression of nature's formal, harmonic structure is attempted in Sir John Davies's "Orchestra, or a Poem on Dancing." His musical vision of cosmic order reveals the heavenly bodies of a Ptolemaic system dancing each to its unique pattern of epicycles. Still, together "their movings make a music that is charming, animated, and courtly."

While Kepler was working on his unique blend of celestial physics and mysticism, he was setting the foundation for Isaac Newton to work out a unified theory of the heavens; meanwhile, a contemporary of Kepler's was challenging some of the basic assumptions about the universe. That man was Galileo Galilei (1564–1642).

Galileo and Kepler had heard of each other, and Kepler sent Galileo a copy of his *Cosmic Mystery*. Galileo replied enthusiastically, complimenting the younger Kepler and stating that he was glad to see another searcher after the truth who was convinced of the validity of the Copernican heliocentric view. He also added that he still taught the Ptolemaic view in his classes, being afraid of the Church's disfavor and of being ridiculed by other scholars. Kepler promptly thanked Galileo for his kind comments but added that he was sorry that Galileo had gone underground with his true beliefs; he shouldn't fear the ignorant, Kepler said. Galileo did not write again to Kepler for another thirteen years. Whether he was offended or frightened, his silence greatly bothered Kepler, especially since Galileo was making discoveries crucial to anyone attempting to construct an accurate picture of the heavens.

Both of Galileo's important discoveries involved the telescope. In 1608 the instrument with a magnification of 7X was revealed at the Frankfurt fair. Soon the Dutch were producing better instruments, and descriptions of them reached Galileo. He constructed his own improved version with a magnification of 10X and turned it toward the heavens. As a result of his investigations he published *The Sidereal Messenger* in 1610. His first discovery revealed that the moon was not a perfectly round sphere, but had craters and mountains on its surface. More than a simple observation, it threw doubt on the perfection of the heavenly spheres which had been, until this point, taken for granted. Next, he pointed out that with his telescope he could see many more stars than could be seen by the naked eye. Some of these were faint, and Galileo concluded that they must be very distant. This contradicted the notion that there were fixed limits to the universe and that the outer periphery was a shell to which the stars were relegated.

The next discovery that Galileo revealed with his telescope was the four moons—which he called planets—of Jupiter. This was an especially important revelation, and it occasioned Galileo's first public defense of the Copernican system. Galileo believed that the main argument of the anti-Copernicans was the impossibility of the moon's motion around the earth while the earth moved around the sun. The example of Jupiter provided empirical evidence of four more bodies that behaved in a manner consonant with the earth's moon as it was described in the Copernican system.

Yet, despite these significant discoveries, Galileo's contribution to astronomy is not quite so great as many believe. This is particularly due to his preoccupation with the physics of falling bodies and with technological projects. While he certainly realized that the telescope was an important tool for astronomy and used it to show that certain archaic views did not correspond to experience, his support of the Copernican system was not so strong as it might have been had he not been so fearful of rebuke or reprisal.

Galileo's reputed cowardice—his recanting of the truth under the pressure of the Inquisition—has haunted his reputation to this day. In 1940 Bertolt Brecht dramatized this piece of history (with implied analogies to Nazi Germany), making Galileo the victim of both fear and egotism. In the play, Galileo recants in order to save himself and eat well. But despite Brecht's intention to criticize the weaknesses of Galileo, the cunning and persistence of the man also survive. He is still a truth seeker able to carry on in secret during unfavorable times and clever enough to bury the information of his work until a more accepting future presents itself.

Reflecting on Brecht's *Galileo*, Muriel Rukeyser's contemporary poem captures the ambivalence of the man with a very personal emphasis on the horror of tainting a new age with betrayal. However, for some poets the age itself became tainted by the new philosophy. This surely is the conviction expressed by John Donne in his "First Anniversary," a poem decrying the loss of an ideal and stable universe.

Indeed, in Galileo's own age we can see how difficult it was for thinkers to commit themselves to new and less comfortable, less acceptable ideas. In the year of Galileo's birth the English playwright Christopher Marlowe was born. The fact that the two men were exact contemporaries permits us to illustrate the uneasy diffusion of current astronomy into the popular imagination. Marlowe's Dr. Faustus calls science "divine astrology." By having the key to its secrets, Faustus hopes to achieve the omnipotence that metaphysics and magic failed to provide him. As part of his notorious pact with the devil, Faustus is permitted to dispute with Mephistopheles the mechanisms of the cosmos. From the devil's answers we can see that his model is still Ptolemaic, regarding the sun and moon as planets revolving about the earth. Mephistopheles is more Galilean in dispelling the crystalline sphere of fixed stars as a myth and in asserting the double motion of planets—which even Faustus regards as common knowledge.

The slow evolution of common knowledge is dramatized even further by a scene from John Milton's *Paradise Lost* (1667). In this poem, written more than half a century later, Adam is still trying to coax the angel Raphael into telling him whether the earth or sun is the center of the universe. Raphael refuses to divulge heavenly secrets, suggesting that admiration rather than investigation is the proper function of man, but we can suspect that this may not have been Milton's own deepest feeling. Milton, who very probably met Galileo in 1638, was fascinated with his "optic glass," which he mentions more than once in *Paradise Lost*. Raphael, flying between heaven and earth, is graced by visual clarity of the shining cosmos greater than "the Glass of Galileo, less assur'd" which observes "imagin'd Lands and Regions in the Moon."[1]

Milton's angel was more ready than Galileo to argue the possibility of lunar inhabitants, which Galileo denied in *Sidereus Nuncius (Starry Messenger)*. But Milton himself, clearly a follower of current science, was able to go only as far as Adam's provocative questioning on the central issue of cosmic structure. Whatever his private views on a heliocentric universe might have been, Milton, again like Marlowe, constructed his lost paradise within the framework of the Ptolemaic system which was still taught in the schools and was more acceptable to his readers. An important exception, however, was Henry More, who by 1647 already had written a long philosophical poem—*The Argument of Democritus Platonissans, or The Infinity of Worlds*—enthusiastically propounding the Copernican world view and the infinity of worlds.

[1] *Paradise Lost*, v. 262–63.

For the most part, popular belief in a sun-centered cosmos waited for Isaac Newton (1642–1727), the great hero of "enlightened" thought. Newton was born the year of Galileo's death. He had already shown promise as a mathematics student at Cambridge when the Great Plague of 1665 forced the students to disperse. Newton went to the country, and it was during this period of forced isolation that he worked out the basics of his celestial mechanics. He did not, however, publish his results until much later, in 1687. The *Philosophie Naturalis Principia Mathematica* made its debut only because Edmund Halley offered to pay the printing costs. (Newton was more interested in the puzzle than in the glory of having solved it.)

Newton's genius was twofold: First, he had the imagination to see how separate threads could be woven into a complete tapestry; second, he alone had the mathematical capacity to complete the task of describing the motion of the heavenly bodies. One theoretical element upon which Newton was able to improve was Galileo's work on inertial motion. For Galileo, all motion was circular (ships moving over the ocean's surface, satellites around planets, etc.). Newton, integrating Descartes's geometry, used the conception of rectilinear motion. It afforded no object a privileged position in space—its space was Euclidean, unbounded. Newton knew that deviations from straight-line motion implied forces of some kind and realized that the same laws describing terrestrial motion might be applied to heavenly bodies.

In working out his theory, Newton was decidedly anti-Greek. He rejected the notion that the stars were fixed and by implication asserted that space is essentially empty. In short, he denied the solidity of the planetary spheres such as Kepler believed in. He attempted in a purely mathematical way to explain the orbits of the planets and thus avoided the various philosophical positions that his predecessors had taken. Newton concluded that the moon is basically a projectile whose centrifugal tendencies are exactly balanced by gravitational pull. Thus planetary motion as well as terrestrial fall is demonstrated to be gravitational. Proceeding from Kepler's three laws, Newton calculated the forces exerted by the sun on the planets and concluded that they are proportional to the masses. Now Newton was able to explain the tides in terms of the gravitational force of the sun and moon on the earth. He even explained the motion of comets in a similar manner, suggesting that they are like planets only with very eccentric elipses around the sun. Further, he realized that the gravitational attraction of two bodies, a function of mass and distance, is reciprocal and falls off as the square of the distance between planets increases. Now that he had a celestial physics, astrology could finally be rejected.

In Newton's age astrology became the victim of satiric attack. Samuel Butler's long anti-Puritan poem "Hudibras" took special care to burlesque at length the activities of a charlatan astrologer, Sidrophel, and his journeyman, who dabble in all manner of trickery. In the section excerpted here the table is turned, and Sidrophel mistakes a kite for a comet. The astrologer's comic musings on the composition of a comet represent a clever parody of current theories like those expressed by More.

Not surprisingly, Butler's opinion was shared by Jonathan Swift, a man generally suspicious of both theoretical and practical science. One of Gulliver's travels took him to the land of the Laputans, who got their names from whores and were abstract mathematicians. Swift described them as having a strong inclination toward astrology and an inability to manage the practical functions of daily life, like preventing their wives from running away.

In spite of all skeptical humorists, mathematics held the most profound place in the new astronomy. Newton's law of gravity is a mathematical expression:

$$G = \frac{m^1 \times m^2}{d^2}$$

It derives in principle from his first two descriptive laws: The first is that a body at rest tends to remain at rest or, if it is in motion, continue in motion in a straight line unless an external force acts upon it; the second is that the force on an object in motion is a product of its mass and acceleration ($F = ma$). In the final synthesis that became the law of gravity through which Newton was able to describe the universe as a coherent whole. Astronomy from this time forward would not offer major revisions; it would merely fill in the details.

In his own rational age Newton's proofs generated a kind of sublime awe that was often expressed in poetry. Marjorie Hope Nicholson's famous book *Newton Demands the Muse* is entirely devoted to these poetic treatments of the man and his theories. Some of the more famous poems reacting to his studies of light are included in our physics section. The poems convey the contemporary attitude that everyone could, in some form, understand Newton's findings and that people would derive great comfort and security from having the universe *at last* explained.

Though the theories remained sure, the security did not survive into the nineteenth century. Indeed, at the end of the eighteenth century, when the universe seemed fixed and fully described, a new planet, Uranus, was discovered on March 13, 1781. The discoverer was William Herschel, the music director at Bath who turned his homemade telescopes on the sky. His strategy was unusual. While professional astronomers were watching sections of the sky that the planets were known to frequent, the more adventurous Herschel watched the uncharted heavens; there he discovered Uranus, which he first believed to be a comet.

As the universe expanded to more remote boundaries, poets of the Romantic age became less convinced that quantified explanations brought them closer to an understanding of man's place in the cosmic scheme. Walt Whitman, for example, in one of his most famous poems, "When I Heard the Learn'd Astronomer," walks out in disgust on the "learn'd astronomer," preferring silence and the mystical night—that is, romantic inner feeling—to all the proofs and figures of objective science. These are feelings shared by Robert Browning, in "My Star," and George Meredith, in "Meditation under Stars," who take their stars to have a private meaning. Emily Dickinson also sets experience and faith against proof. She is troubled by the ambivalence of gravity—atoms fall but planets don't—and yet proofs offer her no consolation in "It Troubled

Me as Once I Was." Paradoxically, she rejects those cosmic problems in order to solve her "larger" (presumably personal) problems, leaving the rest to be answered in heaven. It is interesting that her poem makes some imagistic comparison between atoms and the heavens, for in her century post-Newtonian astronomy made one of its greatest advances by the application of atomic analysis to the heavenly bodies.

Spectroanalysis made possible the analysis of matter—even very distant matter—by measuring the component wavelengths of emitted light (the red and blue that Browning alludes to). The technique began with the discovery that when bodies are heated the atoms are excited and collide. Under such circumstances they emit one or more quanta of light. A particular atom can emit only quanta of certain colors, but when there are also collisions of electrons and atoms, a continuous range of color is emitted. This is why stars emit white light. Using spectroanalysis equipment, it is possible to determine what kinds of atoms are emitting the light. By the end of the last century the discovery of this method enabled scientists to identify rather than speculate about the composition of heavenly bodies.

Since then spectroanalysis has led astronomers to further, more important deductions about the origins and history of these bodies. Stars, in particular, have fired both the scientific and the poetic imaginations, and spectroanalysis has enabled both to understand more fully why "giants" are more luminous than our sun, why "supernovas" explode, and how a "dwarf" is cooled and contracted to glowing ash. The lives of the stars, it has been shown by spectroanalysis, are determined by their composition, by the dominant elements in their spectra. While most stars are remarkably similar in composition, having hydrogen as their common element, they differ because of their masses, which affect their temperatures and consequently their spectra.

Of all the stars, our sun, composed of light helium, has been the greatest focus of "historical" interest—since human life hinges on it. The sun emits an enormous amount of energy; the energy it gives off in one second is more than mankind has consumed in all its history. In the nineteenth century it was thought that perhaps the gravitational attraction of the sun caused meteorites and other bodies to fall onto its surface, and that this was the source of the energy. The correct explanation of the sun's energy involves principles of *thermonuclear* reactions. That word alone is sufficient to epitomize the most powerful feelings generated by our age.

Even before thermonuclear decimation became a possibility on earth, Robinson Jeffers wrote his poem "Nova," about the star that suddenly flourishes in bright explosions, then gradually fades. The poem speculates on the similar fate of our sun. Imagining the inevitable catastrophe on earth, he still finds cause to embrace the "invulnerable beauty of things," a beauty freed from human "consciousness." For Jeffers "consciousness" is the plague of the universe, a contagion man originated in his corner of space. In another poem, "Margrave," Jeffers comes back to Whitman's figure of the "learned astronomer," now versed in spectroanalysis and focused on the "light of remote star-swirls," which Jeffers perceives as fleeing from the taint of human consciousness. Unlike W. H. Auden who, in "Do We Want to Return to the Womb?" argues for a Ptolemaic revival of a man-centered universe, Jeffers awaits the explosion that will rid the uninfected universe of "the bitter weed."

In Jeffers's view egotism and suffering are human messages better annihilated than spread abroad. But for other thinkers and writers the desire to transmit human messages into space in the hope of discovering and communicating with intelligent life-forms on other planets reflects a more humanistic hope. The latest developments in astronomy have given man more sophisticated tools with which to make this quest. The new "star-splitter" is unlike the optical telescope that Robert Frost's determined farmer burns his house to buy. It is based on radio waves that resemble light waves and therefore can be processed like the light from heavenly bodies. After the radio waves have been brought to focus, they are fed into an amplifier that magnifies the signal so that it can be studied. This is the "radio impulse pouring in from Taurus" that Adrienne Rich mentions in "Planetarium." The radio telescope records the total radio brightness of all objects in its field of view. While it gives a much less complete picture than optical astronomy, its virtue is in detecting nebulae that are rather far away from us—as far as 5,000 million light-years.

The sophistication of modern stargazing and the immensity of the cosmos that has been thus far revealed by astronomy are summarized in a poem by May Swenson. It is constructed from a selection of two texts that appeared in the same issue of *The New York Times:* one on telescopes, the other on an excavation at the Vatican. Her "merger" reminds us that both time and space are full of concealed mystery, and that science and religion derive from the common motive of a human need to find one's place. The search for meaning is the universal voyage that Doris Lessing rhapsodically describes in her *Briefing for a Descent into Hell.*

PLATO
(428–348 B.C.)

From *Republic*

BOOK VII

[Socrates is engaged in dialog with Glaucon.]

The spangled heavens should be used as a pattern and with a view to that higher knowledge; their beauty is like the beauty of figures or pictures excellently wrought by the hand of Daedalus, or some other great artist, which we may chance to behold; any geometrician who saw them would appreciate the exquisiteness of their workmanship, but he would never dream of thinking that in them he could find the true equal or the true double, or the truth of any other proportion.

No, he [Glaucon] replied, such an idea would be ridiculous.

And will not a true astronomer have the same feeling

when he looks at the movements of the stars? Will he not think that heaven and the things in heaven are framed by the Creator of them in the most perfect manner? But he will never imagine that the proportions of night and day, or of both to the month, or of the month to the year, or of the stars to these and to one another, and any other things that are material and visible can also be eternal and subject to no deviation—that would be absurd; and it is equally absurd to take so much pains in investigating their exact truth.

I quite agree, though I never thought of this before.

Then, I said, in astronomy, as in geometry, we should employ problems, and let the heavens alone if we would approach the subject in the right way and so make the natural gift of reason to be of any real use.

That, he said, is a work infinitely beyond our present astronomers.

Yes, I said; and there are many other things which must also have a similar extension given to them, if our legislation is to be of any value. But can you tell me of any other suitable study?

No, he said, not without thinking.

Motion, I said, has many forms, and not one only; two of them are obvious enough even to wits no better than ours; and there are others, as I imagine, which may be left to wiser persons.

But what are the two?

There is a second, I said, which is the counterpart of the one already named.

And what may that be?

The second, I said, would seem relatively to the ears to be what the first is to the eyes; for I conceive that as the eyes are designed to look up at the stars, so are the ears to hear harmonious motions; and these are sister sciences—as the Pythagoreans say, and we, Glaucon, agree with them?

Yes, he replied.

But this, I said, is a laborious study, and therefore we had better go and learn of them; and they will tell us whether there are any other applications of these sciences. At the same time, we must not lose sight of our own higher object.

What is that?

There is a perfection which all knowledge ought to reach, and which our pupils ought also to attain, and not fall short of, as I was saying that they did in astronomy. For in the science of harmony, as you probably know, the same thing happens. The teachers of harmony compare the sounds and consonances which are heard only, and their labor, like that of the astronomers, is in vain.

Yes, by heaven! he said; and 'tis as good as a play to hear them talking about their condensed notes, as they call them; they put their ears close alongside of the strings like persons catching a sound from their neighbor's wall—one set of them declaring that they distinguish an intermediate note and have found the least interval which should be the unit of measurement; the others insisting that the two sounds have passed into the same—either party setting their ears before their understanding.

You mean, I said, those gentlemen who tease and torture the strings and rack them on the pegs of the instrument: I might carry on the metaphor and speak after their manner of the blows which the plectrum gives, and make accusations against the strings, both of backwardness and forwardness to sound; but this would be tedious, and therefore I will only say that these are not the men, and that I am referring to the Pythagoreans, of whom I was just proposing to enquire about harmony. For they too are in error, like the astronomers; they investigate the numbers of the harmonies which are heard, but they never attain to problems, that is to say, they never reach the natural harmonies of number, or reflect why some numbers are harmonious and others not.

That, he said, is a thing of more than mortal knowledge.

A thing, I replied, which I would rather call useful; that is, if sought after with a view to the beautiful and good; but if pursued in any other spirit, useless.

Very true, he said.

Now, when all these studies reach the point of intercommunion and connection with one another and come to be considered in their mutual affinities, then, I think, but not till then, will the pursuit of them have a value for our objects; otherwise there is no profit in them.

I suppose so; but you are speaking, Socrates, of a vast work.

What do you mean? I said; the prelude or what? Do you not know that all this is but the prelude to the actual strain which we have to learn? For you surely would not regard the skilled mathematician as a dialectician?

Assuredly not, he said; I have hardly known a mathemetician who was capable of reasoning.

But do you imagine that men who are unable to give and take a reason will have the knowledge which we require of them?

Neither can this be supposed.

And so, Glaucon, I said, we have at last arrived at the hymn of dialectic. This is that strain which is of the intellect only, but which the faculty of sight will nevertheless be found to imitate; for sight, as you may remember, was imagined by us after a while to behold the real animals and stars, and last of all the sun himself. And so with dialectic; when a person starts on the discovery of the absolute by the light of reason only, and without any assistance of sense, and perseveres until by pure intelligence he arrives at the perception of the absolute good, he at last finds himself at the end of the intellectual world, as in the case of sight at the end of the visible.

Exactly, he said.

Then this is the progress which you call dialectic?

True.

From *Timaeus*

Had the earth been a surface only, one mean would have sufficed, but two means are required to unite solid bodies. And as the world was composed of solids, between the elements of fire and earth God placed two other elements of air and water, and arranged them in a continuous proportion—

fire:air::air:water, and air:water::water:earth,

and so put together a visible and palpable heaven, having harmony and friendship in the union of the four elements; and being at unity with itself it was indissoluble except by the hand of the framer. Each of the elements was taken into the universe whole and entire; for he considered that the animal should be perfect and one, leaving no remnants out of which another animal could be created, and should also be free from old age and disease, which are produced by the action of external forces. And as he was to contain all things, he was made in the all-containing form of a sphere, round as from a lathe and every way equidistant from the center, as was natural and suitable to him. He was finished and smooth, having neither eyes nor ears, for there was nothing without him which he could see or hear; and he had no need to carry food to his mouth, nor was there air for him to breathe; and he did not require hands, for there was nothing of which he could take hold, nor feet, with which to walk. All that he did was done rationally in and by himself, which is the most intellectual of motions; but the other six motions were wanting to him; wherefore the universe had no feet or legs.

And so the thought of God made a God in the image of a perfect body, having intercourse with himself and needing no other, but in every part harmonious and self-contained and truly blessed. The soul was first made by him—the elder to rule the younger; not in the order in which our wayward fancy has led us to describe them, but the soul first and afterwards the body.

*   *   *

When the Father who begat the world saw the image which he had made of the eternal Gods moving and living, he rejoiced; and in his joy resolved, since the archetype was eternal, to make the creature eternal as far as this was possible. Wherefore he made an image of eternity which is time, having an uniform motion according to number, parted into months and days and years, and also having greater divisions of past, present, and future. These all apply to becoming in time, and have no meaning in relation to the eternal nature, which ever is and never was or will be; for the unchangeable is never older or younger, and when we say that he "was" or "will be," we are mistaken, for these words are applicable only to becoming, and not to true being; and equally wrong are we in saying that what has become *is* become and that what becomes *is* becoming, and that the non-existence *is* non-existent. . . . These are the forms of time which imitate eternity and move in a circle measured by number.

ARISTOTLE
(384–322 B.C.)

From *On the Heavens*

CHAPTER VIII

That the stars . . . are not rolled along is evident; for that which is rolled along must necessarily be turned round. But what is called the face of the moon is always manifest; so that since it is reasonable to suppose that things which are moved of themselves are moved of their own proper motions, but the stars do not appear to move with these motions; it is evident that they are not moved through themselves. Besides it is irrational to suppose that nature has imparted to them no instrument for the purpose of motion; for nature does nothing casually; nor has she paid attention to animals, and despised things so honourable as the stars; but she seems to have taken away every thing through which they might be able to proceed by themselves, as if such a privation were adapted to them, and because they are very remote from the nature of things which possess instruments of motion. Hence the whole heaven and each of the stars may reasonably appear to be spherical; for a sphere is of all figures the most useful for that motion which is produced in the same place; since it may thus be most rapidly moved, and especially occupy the same place. But such a figure is most useless for an anterior motion; for it is the least similar to those figures that are moved from themselves; since it has nothing depending or prominent like a right-lined figure, but is very remote from the figure of progressive bodies. Since therefore it is necessary that the heaven should be moved with a motion in itself, but the stars should not proceed by themselves, it is reasonable that each of these should be spherical; for thus especially the one will be moved, and the other will be at rest.

CHAPTER IX

From these things also it is evident that to say harmony is generated from the motion of the heavenly bodies, as if concordant sounds were thence produced, is indeed elegantly and subtly said, yet it is not true. For to some it appears to be necessary that a sound should be effected by the motion of such great bodies, since from the bodies with which we are conversant, and which have neither an equal bulk, nor are moved with such celerity as the sun and moon, a sound is produced. To which also they add, that it is impossible an immense sound should not be caused by such a multitude of stars moving with such great celerity. Supposing therefore these things, and also that the celerities have, from the intervals, the ratio of symphonies, they say that a harmonic sound is generated by the revolution of the stars. Since however it appears to be absurd that we should not hear this sound, they say this arises from this sound being familiar to us from our birth; so that it is not manifest from a composition on with a contrary silence. . . . But from the motion of so many and such great bodies, and from the sound pervading to the magnitude which is moved, it is necessary that a multiplied magnitude of sound should arrive hither, and that the strength of the violence should be immense. It is reasonable therefore to suppose that we do not hear any sound from the motion of the celestial orbs, and that bodies do not appear to suffer any violent passion, because no sound is produced.

CLAUDIUS PTOLEMY
(85–165)

From *Tetrabiblos*

CHAPTER I

The studies preliminary to astronomical prognostication, O Syrus! are two: the one, first alike in order and in power, leads to the knowledge of the figurations of the Sun, the Moon, and the stars; and of their relative aspects to each other, and to the earth: the other takes into consideration the changes which their aspects create, by means of their natural properties, in objects under the influence.

The first mentioned study has been already explained in the Syntaxis [*Almagest*] to the utmost practicable extent; for it is complete in itself, and of essential utility even without being blended with the second; to which this treatise will be devoted, and which is equally self-complete. The present work shall, however, be regulated by that due regard for truth which philosophy demands: and since the material quality of the objects acted upon renders them weak and variable, and difficult to be accurately apprehended, no positive or infallible rules (as were given in detailing the first doctrine, which is always governed by the same immutable laws) can be here set forth: while, on the other hand, a due observation of most of those general events, which evidently trace their causes to the Ambient [encompassing space], shall not be omitted.

It is, however, a common practice with the vulgar to slander everything which is difficult of attainment, and surely they who condemn the first of these two studies must be considered totally blind, whatever arguments may be produced in support of those who impugn the second. There are also persons who imagine that whatever they themselves have not been able to acquire, must be utterly beyond the reach of all understanding; while others again will consider as useless any science of which (although they may have been often instructed in it) they have failed to preserve the recollection, owing it its difficulty of retention. In reference to these opinions, therefore, an endeavour shall be made to investigate the extent to which prognostication by astronomy is practicable, as well as serviceable, previously to detailing the particulars of the doctrine.

CHAPTER II
KNOWLEDGE MAY BE ACQUIRED BY ASTRONOMY TO A CERTAIN EXTENT

That a certain power, derived from the aethereal nature, is diffused over and pervades the whole atmosphere of the earth, it is clearly evident to all men. Fire and air, the first of the sublunary elements, are encompassed and altered by the motions of the aether. These elements in their turn encompass all inferior matter, and vary it as they themselves are varied; acting on earth and water, on plants and animals.

The Sun, always acting in connection with the Ambient, contributes to the regulation of all earthly things: not only by the revolution of the seasons does he bring to perfection the embryo of animals, the buds of plants, the spring of waters, and the alteration of bodies, but by his daily progress also he operates other changes in light, heat, moisture, dryness and cold; dependent upon his situation with regard to the zenith.

The Moon, being of all the heavenly bodies the nearest to the Earth, also dispenses much influence; and things animate and inanimate sympathize and vary with her. By the changes of her illumination, rivers swell and are reduced; the tides of the sea are ruled by her risings and settings; and plants and animals are expanded or collapsed, if not entirely at least partially, as she waxes or wanes.

The stars likewise (as well the fixed stars as the planets), in performing their revolutions, produce many impressions on the Ambient. They cause heats, winds, and storms, to the influence of which earthly things are conformably subjected.

And, further, the mutual configurations of all these heavenly bodies, by commingling the influence with which each is separately invested, produce a multiplicity of changes. The power of the Sun however predominates, because it is more generally distributed; the others either co-operate with his power or diminish its effect: the Moon more frequently and more plainly performs this at her conjunction, at her first and last quarter, and at her opposition: the stars act also in a similar purpose, but at longer intervals and more obscurely than the Moon; and their operation principally depends upon the mode of their visibility, their occultation and their declination.[2]

From these premises it follows not only that all bodies, which may be already compounded, are subjected to the motion of the stars, but also that the impregnation and growth of the seeds from which all bodies proceed, are framed and moulded by the quality existing in the Ambient at the time of such impregnation and growth. And it is upon this principle that the more observant husbandmen and shepherds are accustomed, by drawing their inferences from the particular breezes which may happen at seed-time and at the impregnation of their cattle, to form predictions as to the quality of the expected produce. In short, however unlearned in the philosophy of nature, these men can foretell, solely by their previous observation, all the more general and usual effects which result from the plainer and more visible configurations of the Sun, Moon, and stars. It is daily seen that even most illiterate persons, with no other aid than their own experienced observation, are capable of predicting events which may be consequent on the more extended influence of the Sun and the more simple order of the Ambient, and which may not be open to variation by any complex configurations of the Moon and stars towards the Sun. There are, moreover, among the brute creation, animals who evidently form prognostica-

[2]Celestial latitude.

tion, and use this wonderful instinct at the changes of the several seasons of the year, spring, summer, autumn, and winter; and also, at the changes of the wind.

In producing the changes of the seasons, the Sun itself is chiefly the operating and visible cause. There are, however, other events which, although they are not indicated in so simple a manner, but dependent on a slight complication of causes in the Ambient, are also foreknown by persons who have applied their observation to that end. Of this kind, are tempests and gales of wind, produced by certain aspects of the Moon, or the fixed stars, towards the Sun, according to their several courses, and the approach of which is usually seen by mariners. At the same time, prognostication made by persons of this class must be frequently fallacious, owing to their deficiency in science and their consequent inability to give necessary consideration to the time and place, or to the revolutions of the planets; all which circumstances, when exactly defined and understood, certainly tend towards accurate foreknowledge.

When therefore, a thorough knowledge of the motions of the stars, and of the Sun and Moon, shall have been acquired, and when the situation of the place, the time, and all the configurations actually existing at that place and time, shall also be duly known; and such knowledge be yet further improved by an acquaintance with the natures of the heavenly bodies— not of what they are composed, but of the effective influences they possess; as, for instance, that heat is the property of the Sun, and moisture of the Moon, and that other peculiar properties respectively appertain to the rest of them;—when all these qualifications for prescience may be possessed by an individual, there seems no obstacle to deprive him of the insight, offered at once by nature and his own judgment, into the effects arising out of the quality of all the various influences compounded together. So that he will thus be competent to predict the peculiar constitution of the atmosphere in every season, as, for instance, with regard to its greater heat or moisture, or other similar qualities; all which may be foreseen by the visible position or configuration of the stars and the Moon towards the Sun.

Since it is thus clearly practicable, by an accurate knowledge of the points above enumerated, to make predictions concerning the proper quality of the seasons, there also seems no impediment to the formation of similar prognostication concerning the destiny and disposition of every human being. For by the constitution of the Ambient, even at the time of any individual's primary conformation, the general duality of that individual's temperament may be perceived; and the corporeal shape and mental capacity with which the person will be endowed at birth may be pronounced; as well as the favourable and unfavourable events indicated by the state of the Ambient, and liable to attend the individual at certain future periods; since, for instance, an event dependent on one disposition of the Ambient will be advantageous to a particular temperament, and that resulting from another unfavourable and injurious. From these circumstances, and others of similar import, the possibility of prescience is certainly evident. . . .

CHAPTER III
THAT PRESCIENCE IS USEFUL

In exercising prognostication, therefore, strict care must be taken to foretell future events by that natural process only which is admitted in the doctrine here delivered; and, setting aside all vain and unfounded opinions, to predict that, when the existing agency is manifold and great, and of a power impossible to be resisted, the corresponding event which it indicates shall absolutely take place; and also, in other cases, that another event shall not happen when its exciting causes are counteracted by some interposing influence. It is in this manner that experienced physicians, accustomed to the observation of diseases, foresee that some will be inevitably mortal, and that others are susceptible of cure.

. . . Of this, the Ægyptians seem to have been well aware; their discoveries of the great faculties of this science have exceeded those of other nations, and they have in all cases combined the medical art with astronomical prognostication. And, had they been of opinion that all expected events are unalterable and not to be averted, they never would have instituted any propitiations, remedies, and preservatives against the influence of the Ambient, whether present or approaching, general or particular. But, by means of the science called by them Medical Mathematics, they combined with the power of prognostication the concurrent secondary influence arising out of the institutions and courses of nature, as well as the contrary influence which might be procured out of nature's variety; and by means of these they rendered the indicated agency useful and advantageous: since their astronomy pointed out to them the kind of temperament liable to be acted upon, as well as the events about to proceed from the Ambient, and the peculiar influence of those events, while their medical skill made them acquainted with everything suitable or unsuitable to each of the effects procured. And it is by this process that remedies for present and preservations against future disorders are to be acquired: for, without astronomical knowledge, medical aid would be most frequently unavailing; since the same identical remedies are not better calculated for all persons whatsoever, than they are for all diseases whatsoever.

NICOLAUS COPERNICUS
(1473–1543)

From *The Revolution of the Heavenly Spheres*

BOOK ONE

Among the many and varied literary and artistic studies upon which the natural talents of men are nourished, I think that those above all should be embraced and pursued with the most loving care which have to do with things that are very beautiful and very worthy of knowledge. Such studies are those which deal with the godlike circular movements of the world and the course of the stars, their magnitudes, distances, risings, and settings, and the causes of the other appearances in the heavens; and which finally explicate the whole form. For what could be more beautiful than the heavens which contain all beautiful things? Their very names make this clear: *Caelum* (heavens) by naming that which is beautifully carved; and *Mundus* (world), purity and elegance. Many philosophers have called the world a visible god on account of its extraordinary excellence. So if the worth of the arts were measured by the matter with which they deal, this art—which some call astronomy, others astrology, and many of the ancients the consummation of mathematics—would be by far the most outstanding. This art which is as it were the head of all the liberal arts and the one most worthy of a free man leans upon nearly all the other branches of mathematics. Arithmetic, geometry, optics, geodesy, mechanics, and whatever others, all offer themselves in its service. And since a property of all good arts is to draw the mind of man away from the vices and direct it to better things, these arts can do that more plentifully, over and above the unbelievable pleasure of mind [which they furnish]. For who, after applying himself to things which he sees established in the best order and directed by divine ruling, would not through diligent contemplation of them and through a certain habituation be awakened to that which is best and would not wonder at the Artificer of all things, in Whom is all happiness and every good? For the divine Psalmist surely did not say gratuitously that he took pleasure in the workings of God and rejoiced in the works of His hands, unless by means of these things as by some sort of vehicle we are transported to the contemplation of the highest Good.

Now as regards the utility and ornament which they confer upon a commonwealth—to pass over the innumerable advantages they give to private citizens—Plato makes an extremely good point, for in the seventh book of the Laws he says that this study should be pursued in especial, that through it the orderly arrangement of days into months and years and the determination of the times for solemnities and sacrifices should keep the state alive and watchful; and he says that if anyone denies that this study is necessary for a man who is going to take up any of the highest branches of learning, then such a person is thinking foolishly; and he thinks that it is impossible for anyone to become godlike or be called so who has no necessary knowledge of the sun, moon, and the other stars.

However, this more divine than human science, which inquires into the highest things, is not lacking in difficulties. And in particular we see that as regards its principles and assumptions, which the Greeks call "hypotheses," many of those who undertook to deal with them were not in accord and hence did not employ the same methods of calculation. In addition, the courses of the planets and the revolution of the stars cannot be determined by exact calculations and reduced to perfect knowledge unless, through the passage of time and with the help of many prior observations, they can, so to speak, be handed down to posterity. For even if Claud Ptolemy of Alexandria, who stands far in front of all the others on account of his wonderful care and industry, with the help of more than forty years of observations brought this art to such a high point that there seemed nothing left which he had not touched upon; nevertheless we see that very many things are not in accord with the movements which should follow from his doctrine but rather with movements which were discovered later and unknown to him. Whence even Plutarch in speaking of the revolving solar year says, "So far the movement of the stars has overcome the ingenuity of the mathematicians." Now to take the year itself as my example, I believe it is well known how many different opinions there are about it, so that many people have given up hope of making an exact determination of it. Similarly, in the case of the other planets I shall try—with the help of God, without Whom we can do nothing—to make a more detailed inquiry concerning them, since the greater the interval of time between us and the founders of this art—whose discoveries we can compare with the new ones made by us—the more means we have of supporting our own theory. Furthermore, I confess that I shall expound many things differently from my predecessors—although with their aid, for it was they who first opened the road of inquiry into these things.

1. The World Is Spherical

In the beginning we should remark that the world is globe-shaped; whether because this figure is the most perfect of all, as it is an integral whole and needs no joints; or because this figure is the one having the greatest volume and thus is especially suitable for that which is going to comprehend and conserve all things; or even because the separate parts of the world, i.e., the sun, moon, and stars, are viewed under such a form; or because everything in the world tends to be delimited by this form, as is apparent in the case of drops of water and other liquid bodies, when they become delimited of themselves. And so no one would hesitate to say that this form belongs to the heavenly bodies.

2. The Earth Is Spherical Too

The Earth is globe-shaped too, since on every side it rests upon its centre. But it is not perceived straightway to be a perfect sphere, on account of the great height of its mountains and the lowness of its valleys, though they modify its universal roundness to only a very small extent.

That is made clear in this way. For when people journey northward from anywhere, the northern vertex of the axis of daily revolution gradually moves overhead, and the other moves downward to the same extent; and many stars situated to the north are seen not to set, and many to the south are seen not to rise any more. So Italy does not see Canopus, which is visible to Egypt. And Italy sees the last star of Fluvius, which is not visible to this region situated in a more frigid zone. Conversely, for people who travel southward, the second group of stars becomes higher in the sky; while those become lower which for us are high up.

Moreover, the inclinations of the poles have everywhere the same ratio with places at equal distances from the poles of the Earth and that happens in no other figure except the spherical. Whence 'it is manifest that the Earth itself is contained between the vertices and is therefore a globe.

* * *

4. The Movement of the Celestial Bodies Is Regular, Circular, and Everlasting—or Else Compounded of Circular Movements

After this we will recall that the movement of the celestial bodies is circular. For the motion of a sphere is to turn in a circle; by this very act expressing its form, in the most simple body, where beginning and end cannot be discovered or distinguished from one another, while it moves through the same parts in itself.

But there are many movements on account of the multitude of spheres or orbital circles. The most obvious of all is the daily revolution—which the Greeks call νυχθημερον, i.e., having the temporal span of a day and a night. By means of this movement the whole world—with the exception of the Earth—is supposed to be borne from east to west. This movement is taken as the common measure of all movements, since we measure even time itself principally by the number of days.

Next, we see other as it were antagonistic revolutions; i.e., from west to east, on the part of the sun, moon, and the wandering stars. In this way the sun gives us the year, the moon the months—the most common periods of time; and each of the other five planets follows its own cycle. Nevertheless these movements are manifoldly different from the first movement. First, in that they do not revolve around the same poles as the first movement but follow the oblique ecliptic; next, in that they do not seem to move in their circuit regularly. For the sun and moon are caught moving at times more slowly and at times more quickly. And we perceive the five wandering stars sometimes even to retrograde and to come to a stop between these two movements. And though the sun always proceeds straight ahead along its route, they wander in various ways, straying sometimes toward the south, and at other

times toward the north—whence they are called "planets."
Add to this the fact that sometimes they are nearer the
Earth—and are then said to be at their perigee—and at other
times are farther away—and are said to be at their apogee.

We must however confess that these movements are circular or are composed of many circular movements, in that
they maintain these irregularities in accordance with a constant law and with fixed periodic returns: and that could not
take place, if they were not circular. For it is only the circle
which can bring back what is past and over with; and in this
way, for example, the sun by a movment composed of circular
movements brings back to us the inequality of days and
nights and the four seasons of the year. Many movements are
recognized in that movement, since it is impossible that a simple heavenly body should be moved irregularly by a single
sphere. For that would have to take place either on account of
the inconsistency of the motor virtue—whether by reason of
an extrinsic cause or its intrinsic nature—or on account of the
inequality between it and the moved body. But since the mind
shudders at either of these suppositions, and since it is quite
unfitting to suppose that such a state of affairs exists among
things which are established in the best system, it is agreed
that their regular movements appear to us as irregular,
whether on account of their circles having different poles or
even because the Earth is not at the centre of the circles in
which they revolve. And so for us watching from the Earth, it
happens that the transits of the planets, on account of being at
unequal distances from the Earth, appear greater when they
are nearer than when they are farther away, as has been
shown in optics: thus in the case of equal arcs of an orbital
circle which are seen at different distances there will appear to
be unequal movements in equal times. For this reason I think
it necessary above all that we should note carefully what the
relation of the Earth to the heavens is so as not—when we
wish to scrutinize the highest things—to be ignorant of those
which are nearest to us, and so as not—by the same error—to
attribute to the celestial bodies what belongs to the Earth.

5. Does the Earth Have a Circular Movement? And of Its Place

Now that it has been shown that the Earth too has the form of a globe, I think we must see whether or not a movement follows upon its form and what the place of the earth is in the universe. For without doing that it will not be possible to find a sure reason for the movements appearing in the heavens, although there are so many authorities for saying that the Earth rests in the centre of the world that people think the contrary supposition inopinable and even ridiculous; if however we consider the thing attentively, we will see that the question has not been decided and accordingly is by no means to be scorned. For every apparent change in place occurs on account of the movement either of the thing seen or of the spectator, or on account of the necessarily unequal movement of both. For no movement is perceptible relatively to things moved equally in the same directions—I mean relatively to the thing seen and the spectator. Now it is from the Earth that the celestial circuit is beheld and presented to our sight. Therefore, if some movement should belong to the Earth it will appear, in the parts of the universe which are outside, as the same movement but in the opposite direction, as though the things outside were passing over. And the daily revolution appears to carry the whole universe along, with the exception of the Earth and the things around it. And if you admit that the heavens possess none of this movement but that the Earth turns from west to east, you will find—if you make a serious examination—that as regards the apparent rising and setting of the sun, moon, and stars the case is so. And since it is the heavens which contain and embrace all things as the place common to the universe, it will not be clear at once why movement should not be assigned to the contained rather than to the container, to the thing placed rather than to the thing providing the place.

As a matter of fact, the Pythagoreans Herakleides and Ekphantus were of this opinion and so was Hicetas the Syracusan in Cicero; they made the Earth to revolve at the centre of the world. For they believed that the stars set by reason of the interposition of the Earth and that with cessation of that they rose again. Now upon this assumption there follow other things, and a no smaller problem concerning the place of the Earth, though it is taken for granted and believed by nearly all that the Earth is the centre of the world. For if anyone denies that the Earth occupies the midpoint or centre of the world yet does not admit that the distance [between the two] is great enough to be compared with [the distance to] the sphere of the fixed stars but is considerable and quite apparent in relation to the orbital circles of the sun and the planets; and if for that reason he thought that their movements appeared irregular because they are organized around a different centre from the centre of the Earth, he might perhaps be able to bring forward a perfectly sound reason for movement which appears irregular. For the fact that the wandering stars are seen to be sometimes nearer the Earth and at other times farther away necessarily argues that the centre of the Earth is not the centre of their circles. It is not yet clear whether the Earth draws near to them and moves away or they draw near to the Earth and move away.

And so it would not be very surprising if someone attributed some other movement to the Earth in addition to the daily revolution. As a matter of fact, Philolaus the Pythagorean—no ordinary mathematician, whom Plato's biographers say Plato went to Italy for the sake of seeing—is supposed to have held that the Earth moved in a circle and wandered in some other movements and was one of the planets.

Many however have believed that they could show by geometrical reasoning that the Earth is in the middle of the world; that it has the proportionality of a point in relation to the immensity of the heavens, occupies the central position, and for this reason is immovable, because, when the universe moves, the centre remains unmoved and the things which are closest to the centre are moved the most slowly.

* * *

8. Answer to the Aforesaid Reasons and Their Inadequacy

. . . It seems rather absurd to ascribe movement to the container or to that which provides the place and not rather to that which is contained and has a place, i.e., the Earth. And lastly, since it is clear that the wandering stars are sometimes nearer and sometimes farther away from the Earth, then the movement of one and the same body around the centre—and they mean the centre of the Earth—will be both away from the centre and toward the centre. Therefore it is necessary that movement around the centre should be taken more generally; and it should be enough if each movement is in accord with its own centre. You see therefore that for all these reasons it is more probable that the Earth moves than that it is at rest—especially in the case of the daily revolution, as it is the Earth's very own. . . .

9. Whether Many Movements Can Be Attributed to the Earth and Concerning the Centre of the World

Therefore, since nothing hinders the mobility of the Earth, I think we should now see whether more than one movement belongs to it, so that it can be regarded as one of the wandering stars. For the apparent irregular movement of the planets and their variable distances from the Earth—which cannot be understood as occurring in circles homocentric with the Earth—make it clear that the Earth is not the centre of their circular movements. Therefore, since there are many centres, it is not foolhardy to doubt whether the centre of gravity of the Earth rather than some other is the centre of the world. I myself think that gravity or heaviness is nothing except a certain natural appetency implanted in the parts by divine providence of the universal Artisan, in order that they should unite with one another in their oneness and wholeness and come together in the form of a globe. It is believable that this affect is present in the sun, moon, and the other bright planets and that through its efficacy they remain in the spherical figure in which they are visible, though they nevertheless accomplish their circular movements in many different ways. Therefore if the Earth too possesses movements different from the one around its centre, then they will necessarily be movements which similarly appear on the outside in the many bodies; and we find the yearly revolution among these movements. For if the annual revolution were changed from being solar to being terrestrial, and immobility were granted to the sun, the risings and settings of the signs and of the fixed stars—whereby they become morning or evening stars—will appear in the same way; and it will be seen that the stoppings, retrogressions, and progressions of the wandering stars are not their own, but are a movement of the Earth and that they borrow the appearances of this movement. Lastly, the sun will be regarded as occupying the centre of the world. And the ratio of order in which these bodies succeed one another and the harmony of the whole world teaches us their truth, if only—as they say—we would look at the thing with both eyes.

JOHANNES KEPLER

(1571–1630)

From *Harmonies of the World*

6. IN THE EXTREME PLANETARY MOVEMENTS THE MUSICAL MODES OR TONES HAVE SOMEHOW BEEN EXPRESSED

This follows from the aforesaid and there is no need of many words; for the single planets somehow mark the pitches of the system with their perihelial[3] movement, in so far as it has been appointed to the single planets to traverse a certain fixed interval in the musical scale comprehended by the definite notes of it or the pitches of the system, and beginning at that note or pitch of each planet which in the preceding chapter fell to the aphelial[4] movement of that planet: G to Saturn and

[3]The movement of a planet toward its orbital point closest to the sun.
[4]The opposing movement of a planet toward its orbital point furthest from the sun.

the Earth; *B* to Jupiter, which can be transposed higher to *G*; *F* sharp to Mars; *E* to Venus; *A* to Mercury in the higher octave. See the single movements in the familiar terms of notes. They do not form articulately to the intermediate positions, which you see here filled by notes, as they do the extremes, because they struggle from one extreme to the opposite not by leaps and intervals but by a continuum of tunings and actually traverse all the means (which are potentially infinite)—which cannot be expressed by me in any other way than by a continuous series of intermediate notes. Venus remains approximately in unison and does not equal even the least of the concordant intervals in the difference of its tension.

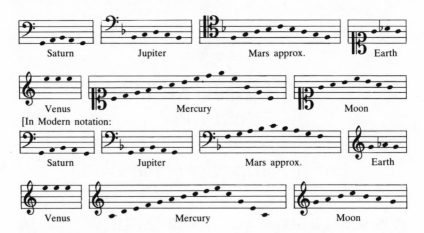

But the signature of two accidentals (flats) in a common staff and the formation of the skeletal outline of the octave by the inclusion of a definite concordant interval are a certain first beginning of the distinction of Tones or Modes [*modorum*]. Therefore the musical Modes have been distributed among the planets. But I know that for the formation and determination of distinct Modes many things are requisite, which belong to human song, as containing (a) distinct [order of] intervals; and so I have used the word *somehow*.

But the harmonist will be free to choose his opinion as to which Mode each planet expresses as its own, since the extremes have been assigned to it here. From among the familiar Modes, I should give to Saturn the Seventh or Eighth, because if you place its key-note at *G,* the perihelial movement ascends to *B;* to Jupiter, the First or Second Mode, because its aphelial movement has been fitted to *G* and its perihelial movement arrives at *B* flat; to Mars, the Fifth or Sixth Mode, not only because Mars comprehends approximately the perfect fifth, which interval is common to all the Modes, but principally because when it is reduced with the others to a common system, it attains *C* with its perihelial movement and touches *F* with its aphelial, which is the key-note of the Fifth or Sixth Mode or Tone; I should give the Third or Fourth Mode to the Earth, because its movement revolves within a semitone, while the first interval of those Modes is a semitone; but to Mercury will belong indifferently all the Modes or Tones on account of the greatness of its range; to Venus, clearly none on account of the smallness of its range; but on account of the common system the Third and Fourth Mode, because with reference to the other planets it occupies *E.* (The Earth sings *MI, FA, MI* so that you may infer even from the syllables that in this our domicile *MI*sery and *FA*mine obtain.)

WILLIAM SHAKESPEARE

(1564–1616)

Sonnet 14

Not from the stars do I my judgement pluck,
And yet methinks I have astronomy,
But not to tell of good or evil luck,
Of plagues, of dearths, or seasons' quality.
Nor can I fortune to brief minutes tell,
Pointing to each his thunder, rain, and wind,
Or say with princes if it shall go well

By oft predict that I in heaven find.
But from thine eyes my knowledge I derive,
And, constant stars, in them I read such art
As truth and beauty shall together thrive,
If from thyself to store thou wouldst convert.
　　Or else of thee this I prognosticate:
　　Thy end is truth's and beauty's doom and date.

ROBERT HERRICK
(1591–1674)

To Music.　A Song

Music, thou queen of heaven, care-charming spell,
　　That strik'st a stillness into hell:
Thou that tam'st tigers, and fierce storms that rise,
　　With thy soul-melting lullabies:
Fall down, down, down, from those thy chiming spheres,
　　To charm our souls, as thou enchant'st our ears.

SIR JOHN DAVIES
(1569–1626)

From "Orchestra or, a Poem on Dancing"

'Dancing, bright lady, then began to be
When the first seeds whereof the world did spring,
The fire air earth and water, did agree
By Love's persuasion, nature's mighty king,
To leave their first discorded combating
And in dance such measure to observe
As all the world their motion should preserve.

'Since when they still are carried in a round,
And changing come one in another's place;
Yet do they neither mingle nor confound,
But every one doth keep the bounded space
Wherein the dance doth bid it turn or trace.
This wondrous miracle did Love devise,
For dancing is love's proper exercise.

'Like this he framed the gods' eternal bower,
And of a shapeless and confusèd mass,
By his through-piercing and digesting power,
The turning vault of heaven formèd was,
Whose starry wheels he hath so made to pass,
As that their movings do a music frame,
And they themselves still dance unto the same.

*   *   *

"'What if to you these sparks disordered seem,
As if by chance they had been scattered there?
The gods a solemn measure do it deem
And see a just proportion everywhere,
And know the points when first their movings were,
To which first points when all return again,
The axletree of heaven shall break in twain.

"'Under that spangled sky five wandering flames,
Besides the king of day and queen of night,
Are wheel'd around, all in their sundry frames,
And all in sundry measures do delight;
Yet altogether keep no measure right;
For by itself each doth itself advance,
And by itself each doth a galliard dance.

"'Venus, the mother of that bastard Love
Which doth usurp the world's great marshal's name,
Just with the sun her dainty feet doth move,
And unto him doth all her gestures frame;
Now after, now afore, the flattering dame
With divers cunning passages doth err,
Still him respecting that respects not her.

"'For that brave sun, the father of the day,
Doth love this earth, the mother of the night,
And like a reveller in rich array
Doth dance his galliard in his leman's sight,
Both back and forth and sideways passing light.
His gallant grace doth so the gods amaze
That all stand still and at his beauty gaze.

"'But see the earth when she approacheth near,
How she for joy doth spring and sweetly smile;
But see again her sad and heavy cheer
When changing places he retires awhile.
But those black clouds he shortly will exile,
And make them all before his presence fly
As mists consumed before his cheerful eye.

"'Who doth not see the measure of the moon?
Which thirteen times she danceth every year,
And ends her pavan thirteen times as soon
As doth her brother, of whose golden hair
She borroweth part and proudly doth it wear.
Then doth she coyly turn her face aside,
That half her cheek is scarce sometimes descried.

"'Next her, the pure, subtle, and cleansing fire
Is swiftly carried in a circle even,
Though Vulcan be pronounced by many a liar
The only halting god that dwells in heaven;
But that foul name may be more fitly given
To your false fire, that far from heaven is fall,
And doth consume, waste, spoil, disorder all.

"'And now behold your tender nurse, the air,
And common neighbour that aye runs around;
How many pictures and impressions fair
Within her empty regions are there found
Which to your senses dancing do propound?
For what are breath, speech, echoes, music, winds,
But dancings of the air in sundry kinds?

GALILEO GALILEI
(1564–1642)

From *The Starry Messenger*

TO THE MOST SERENE COSMO DE' MEDICI, THE SECOND, FOURTH GRAND-DUKE OF TUSCANY

There is certainly something very noble and large-minded in
the intention of those who have endeavoured to protect from
envy the noble achievements of distinguished men, and to

rescue their names, worthy of immortality, from oblivion and
decay. This desire has given us the lineaments of famous
men, sculptured in marble, or fashioned in bronze, as a me-
morial of them to future ages; to the same feeling we owe the
erection of statues, both ordinary and equestrian; hence, as
the poet [Propertius] says, has originated expenditure, mounting to the stars, upon columns and pyramids; with this desire,
lastly, cities have been built, and distinguished by the names
of those men, whom the gratitude of posterity thought worthy of being handed down to all ages. For the state of the
human mind is such, that unless it be continually stirred by
the counterparts of matters, obtruding themselves upon it
from without, all recollection of the matters easily passes
away from it.

But others, having regard for more stable and more lasting monuments, secured the eternity of the fame of great men by placing it under the protection, not of marble or bronze, but of the Muses' guardianship and the imperishable monuments of literature. But why do I mention these things, as if human wit, content with these regions, did not dare to advance further; whereas, since she well understood that all human monuments do not perish at last by violence, by weather, or by age, she took a wider view, and invented more imperishable signs, over which destroying Time and envious Age could claim no rights; so, betaking herself to the sky, she inscribed on the well-known orbs of the brightest stars—those everlasting orbs—the names of those who, for eminent and god-like deeds, were accounted worthy to enjoy an eternity in company with the stars. Wherefore the fame of Jupiter, Mars, Mercury, Hercules, and the rest of the heroes by whose names the stars are called, will not fade until the extinction of the splendour of the constellations themselves.

But this invention of human shrewdness, so particularly noble and admirable, has gone out of date ages ago, inasmuch as primeval heroes are in possesion of those bright abodes, and keep them by a sort of right; into whose company the affection of Augustus in vain attempted to introduce Julius Cæsar; for when he wished that the name of the Julian constellation should be given to a star, which appeared in his time, one of those which the Greeks and the Latins alike name, from their hair-like tails, comets, it vanished in a short time and mocked his too eager hope. But we are able to read the heavens for your highness, most Serene Prince, far more truly and more happily, for scarcely have the immortal graces of your mind begun to shine on earth, when bright stars present themselves in the heavens, like tongues to tell and celebrate your most surpassing virtues to all time. Behold therefore, four stars reserved for your famous name, and those not belonging to the common and less conspicuous multitude of fixed stars, but in the bright ranks of the planets— four stars which, moving differently from each other, round the planet Jupiter, the most glorious of all the planets, as if they were his own children, accomplish the course of their orbits with marvelous velocity, while all the while with one accord they complete all together mighty revolutions every ten years round the centre of the universe, that is, round the Sun.

But the Maker of the Stars himself seems to direct me by clear reasons to assign these new planets to the famous name of your highness in preference to all others. For just as these stars, like children worthy of their sire, never leave the side of Jupiter by any appreciable distance, so who does not know that clemency, kindness of heart, gentleness of manners, splendour of royal blood, nobleness in public functions, wide extent of influence and power over others, all of which have fixed their common abode and seat in your highness—who, I say, does not know that all these qualities, according to the providence of God, from whom all good things do come, emanate from the benign star of Jupiter? Jupiter, Jupiter, I maintain, at the instant of the birth of your highness having at length emerged from the turbid mists of the horizon, and being in possession of the middle quarter of the heavens, and illuminating the eastern angle, from his own royal house, from that exalted throne, looked out upon your most happy birth, and poured forth into a most pure atmosphere all the brightness of his majesty, in order that your tender body and your mind—though that was already adorned by God with still more splendid graces—might imbibe with your first breath the whole of that influence and power. But why should I use only plausible arguments when I can almost absolutely demonstrate my conclusion? It was the will of Almighty God that I should be judged by your most serene parents not unworthy to be employed in teaching your highness mathematics, which duty I discharged, during the four years just passed, at that time of the year when it is customary to take a relaxation from severer studies. Wherefore, since it evidently fell to my lot by God's will, to serve your highness, and so to receive the rays of your surpassing clemency and beneficence in a position near your person, what wonder is it if you have so warmed my heart that it thinks about scarcely anything else day and night, but how I, who am indeed your subject not only by inclination, but also by my very birth and lineage, may be known to be most anxious for your glory, and most grateful to you? And so, inasmuch as under your patronage, most serene Cosmo, I have discovered these stars, which were unknown to all astronomers before me, I have, with very good right, determined to designate them with the most august name of your family. And as I was the first to investigate them, who can rightly blame me if I give them a name, and call them *the Medicean Stars*, hoping that as much consideration may accrue to these stars from this title, as other stars have brought other heroes? For not to speak of your most serene ancestors, to whose everlasting glory the monuments of all history bear witness, your virtue alone, most mighty sire, can confer on those stars an immortal name; for who can doubt that you will not only maintain and preserve the expectations, high though they may be, about yourself, which you have aroused by the very happy beginning of your government, but that you will also far surpass them, so that when you have conquered others like yourself, you may still vie with yourself, and become day by day greater than yourself and your greatness?

Accept, then, most clement Prince, this addition to the glory of your family, reserved by the stars for you; and may you enjoy for many years those good blessings, which are sent to you not so much from the stars as from God, the Maker and Governor of the stars.

Your Highness's most devoted servant,

GALILEO GALILEI

PADUA, *March 12, 1610.*

THE ASTRONOMICAL MESSENGER

Containing and setting forth Observations lately made with the aid of a newly invented Telescope respecting the Moon's Surface, the Milky Way, Nebulous Stars, an innumerable multitude of fixed stars, and also respecting Four Planets never before seen which have been named

The Cosmian Stars

In the present small treatise I set forth some matters of great interest for all observers of natural phenomena to look at and consider. They are of great interest, I think, first, from their intrinsic excellence; secondly, from their absolute novelty; and lastly, also on account of the instrument by the aid of which they have been presented to my apprehension.

The number of the Fixed Stars which observers have been able to see without artificial powers of sight up to this day can be counted. It is therefore decidedly a great feat to add to their number, and to set distinctly before the eyes other stars in myriads, which have never been seen before, and which surpass the old, previously known, stars in number more than ten times.

Again, it is a most beautiful and delightful sight to behold the body of the Moon, which is distant from us nearly sixty *semi*-diameters of the Earth, as near as if it was at a distance of only two of the same measures; so that the diameter of this same Moon appears about thirty times larger, its surface about nine hundred times, and its solid mass nearly 27,000 times larger than when it is viewed only with the naked eye; and consequently any one may know with certainty that is due to the use of our senses, that the Moon certainly does not possess a smooth and polished surface, but one rough and uneven, and, just like the face of the Earth itself, is everywhere full of vast protuberances, deep chasms, and sinuosities.

Then to have got rid of disputes about the Galaxy or Milky Way, and to have made its nature clear to the very senses, not to say to the understanding, seems by no means a matter which ought to be considered of slight importance. In addition to this, to point out, as with one's finger, the nature of those stars which every one of the astronomers up to this time has called *nebulous,* and to demonstrate that it is very different from what has hitherto been believed, will be pleasant, and very fine. But that which will excite the greatest astonishment by far, and which indeed especially moved me to call the attention of all astronomers and philosophers, is this, namely, that I have discovered four planets, neither known nor observed by any one of the astronomers before my time, which have their orbits round a certain bright star, one of those previously known, like Venus and Mercury round the Sun, and are sometimes in front of it, sometimes behind it, though they never depart from it beyond certain limits. All which facts were discovered and observed a few days ago by the help of a telescope devised by me, through God's grace first enlightening my mind.

Perchance other discoveries still more excellent will be made from time to time by me or by other observers, with the assistance of a similar instrument, so I will first briefly record its shape and preparation, as well as the occasion of its being devised, and then I will give an account of the observations made by me.

About ten months ago a report reached my ears that a Dutchman had constructed a telescope, by the aid of which visible objects, although at a great distance from the eye of the observer, were seen distinctly as if near; and some proofs of its most wonderful performances were reported, which some gave credence to, but others contradicted. A few days after, I received confirmation of the report in a letter written from Paris by a noble Frenchman, Jaques Badovere, which finally determined me to give myself up first to inquire into the principle of the telescope, and then to consider the means by which I might compass the invention of a similar instrument, which a little while after I succeeded in doing through deep study of the theory of Refraction; and I prepared a tube, at first of lead, in the ends of which I fitted two glass lenses, both plane on one side, but on the other side one spherically convex and the other concave. Then bringing my eye to the concave lens I saw objects satisfactorily large and near, for they appeared one-third of the distance off and nine times larger than when they are seen with the natural eye alone. I shortly afterwards constructed another telescope with more nicety, which magnified objects more than sixty times. At length, by sparing neither labour nor expense, I succeeded in constructing for myself an instrument so superior that objects seen through it appear magnified nearly a thousand times, and more than thirty times nearer than if viewed by the natural powers of sight alone.

*   *   *

Now let me review the observations made by me during the two months just past, again inviting the attention of all who are eager for true philosophy to the beginnings which led to the sight of most important phenomena.

Let me speak first of the surface of the Moon, which is turned towards us. For the sake of being understood more easily, I distinguish two parts in it, which I call respectively the brighter and the darker. The brighter part seems to surround and pervade the whole hemisphere; but the darker part, like a sort of cloud, discolours the Moon's surface and makes it appear covered with spots. Now these spots, as they are somewhat dark and of considerable size, are plain to every one, and every age has seen them, wherefore I shall call them *great* or *ancient* spots, to distinguish them from other spots, smaller in size, but so thickly scattered that they sprinkle the whole surface of the Moon, but especially the brighter portion of it. These spots have never been observed by any one before me; and from my observation of them, often repeated, I have been led to the opinion which I have expressed, namely, that I feel sure that the surface of the Moon is not perfectly smooth, free from inequalities and exactly spherical, as a large school of philosophers considers with regard to the Moon and the other heavenly bodies, but that, on the contrary, it is full of inequalities, uneven, full of hollows and protuberances, just like the surface of the Earth itself, which is varied everywhere by lofty mountains and deep valleys.

The appearances from which we may gather these conclusions are of the following nature:—On the fourth or fifth day after new-moon, when the Moon presents itself to us with bright horns, the boundary which divides the part in shadow from the enlightened part does not extend continuously in an ellipse, as would happen in the case of a perfectly spherical body, but it is marked out by an irregular, uneven, and very wavy line . . . for several bright excrescences, as they may be called, extend beyond the boundary of light and shadow into the dark part, and on the other hand pieces of shadow encroach upon the light:—nay, even a great quantity of small blackish spots, altogether separated from the dark part, sprinkle everywhere almost the whole space which is at the time flooded with the Sun's light, with the exception of that part alone which is occupied by the great and ancient spots. I have noticed that the small spots just mentioned have this common characteristic always and in every case, that they have the dark part towards the Sun's position, and on the side away from the Sun they have brighter boundaries, as if they were crowned with shining summits. Now we have an appearance quite similar on the Earth about sunrise, when we behold the valleys, not yet flooded with light, but the mountains surrounding them on the side opposite to the Sun already ablaze with the splendour of his beams; and just as the shadows in the hollows of the Earth diminish in size as the Sun rises higher, so also these spots on the Moon lose their blackness as the illuminated part grows larger and larger. Again, not only are the boundaries of light and shadow in the Moon seen to be uneven and sinuous, but—and this produces still greater astonishment—there appear very many bright points within the darkened portion of the Moon, altogether divided and broken off from the illuminated tract, and separated from it by no inconsiderable interval, which, after a little while, gradually increase in size and brightness, and after an hour or two become joined on to the rest of the bright portion, now become somewhat larger; but in the meantime others, one here and another there, shooting up as if growing, are lighted up within the shaded portion, increase in size, and at last are linked on to the same luminous surface, now still more extended. . . . Now, is it not the case on the Earth before sunrise, that while the level plain is still in shadow, the peaks of the most lofty mountains are illuminated by the Sun's rays? After a little while does not the light spread further, while the middle and larger parts of those mountains are becoming illuminated; and at length, when the Sun has risen, do not the illuminated parts of the plains and hills join together? The grandeur, however, of such prominences and depressions in the Moon seems to surpass both in magnitude and extent the ruggedness of the Earth's surface, as I shall hereafter show. And here I cannot refrain from mentioning what a remarkable spectacle I observed while the Moon was rapidly approaching her first quarter. . . . A protuberance of the shadow, of great size, indented the illuminated part in the neighbourhood of the lower cusp; and when I had observed this indentation longer, and had seen that it was dark throughout, at length, after about two hours, a bright peak began to arise a little below the middle of the depression; this by degrees increased, and presented a triangular shape, but was as yet quite detached and separated from the illuminated surface. Soon around it three other small points began to shine, until, when the Moon was about to set, that triangular figure, having now extended and widened, began to be connected with the rest of the illuminated part, and, still girt with the three bright peaks already mentioned, suddenly burst into the indentation of shadow like a vast promontory of light.

ROBERT BURTON
(1577–1640)

From *The Anatomy of Melancholy*

'Tis true, they say, according to optic principles, the visible appearances of the planets do so indeed answer to their magnitudes and orbs and come nearest to mathematical observations and precedent calculations; there is no repugnancy to physical axioms because no penetration of orbs. But then between the sphere of Saturn and the firmament there is such an incredible and vast space of distance—7,000,000 semidiameters of the earth, as Tycho calculates—void of stars. And besides, they do so enhance the bigness of the stars, enlarge their circuit to solve those ordinary objections of parallaxes and retrogradations of the fixed stars, that alteration of the poles, elevation in several places or latitude of cities here on earth; for, say they, if a man's eye were in the firmament, he should not at all discern that great annual motion on the earth, but it would still appear *punctum indivisible* and seem to

be fixed in one place, of the same bigness, that it is quite opposite to reason, to natural philosophy, and all out as absurd
as disproportional, so some will as prodigious as that of the
sun's swift motion of heavens. But *hoc posito*, to grant this
their tenet of the earth's motion, if the earth move, it is a
planet and shines to them in the moon and to the other plane-

tary inhabitants as the moon and they do to us upon the earth. But shine she doth, as Galileo, Kepler, and others prove, and then *per consequens* the rest of the planets are inhabited as well as the moon, which he grants in his dissertation with Galileo's *Nuncius Sidereus* [*Starry Messenger*], "that there be Jovial and Saturn inhabitants," etc., and those several planets have their several moons about them as the earth hath hers, as Galileus hath already evinced by his glasses, four about Jupiter, two about Saturn (Sitius the Florentine, Fortunius Licetus, and Jul. Caesar la Galla cavil at it), yet Kepler, the emperor's mathematician, confirms out of this experience that he saw as much by the same help and more about Mars, Venus; and the rest they hope to find out peradventure even amongst the fixed stars, which Brunus and Brutius have already averred. Then I say, the earth and they be planets alike. . . . If our world be small in respect, why may we not suppose a plurality of worlds, those infinite stars visible in the firmament to be so many suns with particular fixed centers, to have likewise their subordinate planets, as the sun hath his dancing still round him?

*   *   *

But hoo! I am now gone quite out of sight, I am almost giddy with roving about. I could have ranged farther yet; but I am an infant and not able to dive into these profundities or sound these depths; not able to understand, much less to discuss. I leave the contemplation of these things to stronger wits, that have better ability and happier leisure to wade into such philosophical mysteries. For put case I were as able as willing, yet what can one man do? . . . When God sees his time, he will reveal these mysteries to mortal men and show that to some few at last which he hath concealed so long.

[Pt. 2, sect. 2, numb. 3]

JOHN DONNE
(1572–1631)

From "The First Anniversary"

And new Philosophy calls all in doubt,
The Element of fire is quite put out;
The Sun is lost, and th'earth, and no mans wit
Can well direct him where to looke for it.
And freely men confesse that this world's spent,
When in the Planets, and the Firmament
They seeke so many new; then see that this
Is crumbled out againe to his Atomies.
'Tis all in peeces, all cohaerence gone.

*   *   *

We thinke the heavens enjoy their Sphericall,
Their round proportion embracing all.
But yet their various and perplexed course,
Observ'd in divers ages, doth enforce
Men to finde out so many Eccentrique parts,
Such divers downe-right lines, such overthwarts,
As disproportion that pure forme: It teares
The Firmament in eight and forty sheires,
And in these Constellations then arise
New starres, and old doe vanish from our eyes:
As though heav'n suffered earthquakes, peace or war.
When new Towers rise, and old demolish't are.
They have impal'd within a Zodiake
The free-borne Sun, and keepe twelve Signes awake
To watch his steps; the Goat and Crab controule,

And fright him backe, who else to either Pole
(Did not these Tropiques fetter him) might runne:
For his course is not round; nor can the Sunne
Perfit a Circle, or maintaine his way
One inch direct; but where he rose to-day
He comes no more, but with a couzening line,
Steales by that point, and so is Serpentine:
And seeming weary with his reeling thus,
He means to sleepe, being now falne nearer us.
So, of the Starres which boast that they doe runne
In Circles still, none ends where he begun.
All their proportion's lame, it sinkes, it swels.
For of Meridians, and Parallels,
Man hath weav'd out a net, and this net throwne
Upon the Heavens, and now they are his owne.

CHRISTOPHER MARLOWE

(1564–93)

From *Dr. Faustus*

FAUSTUS: Divinity adieu!
These metaphysics of magicians,
And necromantic books are heavenly;
Lines, circles, signs, letters, and characters!
Ay, these are those that Faustus most desires.
O what a world of profit and delight,
Of power, honour, of omnipotence,
Is promised to the studious artisan!
All things that move between the quiet poles
Shall be at my command: emperors and kings,
Are but obeyed in their several provinces,

Nor can they raise the wind, or rend the clouds;
But his dominion that exceeds in this
Stretcheth as far as doth the mind of man:
A sound magician is a mighty god;
Here, Faustus, try thy brains to gain a deity.   *[I,i. 48–63]*

*   *   *

FAUSTUS: I am resolved! Faustus shall not repent.
Come, Mephistopheles, let us dispute again,
And reason of divine astrology.
Speak, are there many spheres above the moon?
Are all celestial bodies but one globe,
As is the substance of this centric earth?
   MEPHISTOPHELES: As are the elements, such are the
     heavens,
Even from the moon unto the empyreal orb,
Mutually folded in each other's spheres,
And jointly move upon one axletree,
Whose termine is termed the world's wide pole.
Nor are the names of Saturn, Mars, or Jupiter,
Feigned, but are erring stars.
   FAUSTUS: But have they one motion, both *situ et tempore?*[5]
   MEPHISTOPHELES: All move from east to west in four and
     twenty hours, upon the poles of the world, but differ in
     their motions upon the poles of the zodiac.
   FAUSTUS: These slender questions Wagner [Faustus'
     servant] can decide!
Hath Mephistopheles no greater skill?
Who knows not the double motion of the planets?
That the first is finished by a natural day, the second thus:
Saturn in thirty years; Jupiter in twelve; Mars in four; the
     Sun
Venus and Mercury in a year; the Moon in twenty-eight days.
     These are freshmen's suppositions. But tell me, hath
     every sphere a dominion or intelligence?[6]
   MEPHISTOPHELES: Ay.
   FAUSTUS: How many heavens, or spheres, are there?

[5]"In direction and time."
[6]"Orders of angels."

MEPHISTOPHELES: None: the seven planets, the firmament, and the empyreal heaven.

FAUSTUS: But is there not *Coelum igneum? et cristallinum?*[7]

MEPHISTOPHELES: No, Faustus, they be but fables.

FAUSTUS: Resolve me then in this one question: why are not conjunctions,

oppositions, aspects, eclipses, all at one time,

but in some years we have more, in some less?

MEPHISTOPHELES: *Per naequalem motum, respectu totius.*[8]

FAUSTUS: Well, I am answered. Now tell me who made the world?

MEPHISTOPHELES: I will not.   *[II, ii. 31–67]*

JOHN MILTON
(1608–74)

From *Paradise Lost*

[ADAM IS SPEAKING]

. . . something yet of doubt remains,
Which only thy solution can resolve.
When I behold this goodly Frame, this World,
Of Heav'n and Earth consisting, and compute
Their magnitudes, this Earth, a spot, a grain,
An Atom, with the Firmament compar'd
And all her number'd Stars, that seem to roll
Spaces incomprehensible (for such
Thir distance argues and their swift return
Diurnal) merely to officiate light

[7]"Heavens of fire and crystals."
[8]"By an irregular motion with respect to the whole."

Round this opacous Earth, this punctual spot,
One day and night; in all their vast survey
Useless besides; reasoning, I oft admire,
How Nature, wise and frugal could commit
Such disproportions, with superfluous hand
So many nobler Bodies to create,
Greater so manifold to this one use,
For aught appears, and on their Orbs impose
Such restless revolution day by day
Repeated, while the sedentary Earth,
That better might with far less compass move,
Serv'd by more noble than herself, attains
Her end without least motion, and receives,
As tribute, such a sumless journey brought
Of incorporeal speed, her warmth and light;
Speed, to describe whose swiftness Number
    fails.   *[VIII, 13–38]*

[RAPHAEL, GOD'S ANGEL, REPLIES]

To ask or search I blame thee not, for Heav'n
Is as the Book of God before thee set,
Wherein to read his wond'rous works, and learn
His Seasons, Hours, or Days, or Months, or Years:
This to attain, whether Heav'n move or Earth,
Imports not, if thou reck'n right; the rest
From Man or Angel the great Architect
Did wisely to conceal, and not divulge
His secrets, to be scann'd by them who ought
Rather admire; or if they list to try
Conjecture, he his Fabric of the Heav'ns
Hath left to their disputes, perhaps to move
His laughter at their quaint Opinions wide
Hereafter, when they come to model Heav'n
And calculate the Stars, how they will wield
The mighty frame, how build, unbuild, contrive
To save appearances, how gird the Sphere
With Centric and Eccentric scribbl'd o'er,
Cycle and Epicycle, Orb in Orb.

Already by thy reasoning this I guess
Who art to lead thy offspring, and supposest
That bodies bright and greater should not serve
The less not bright, nor Heav'n such journeys run,
Earth sitting still, when she alone receives
The benefit: consider, first, that Great
Or Bright infers not Excellence: The Earth
Though, in comparison of Heav'n, so small,
Nor glistering, may of solid good contain
More plenty than the Sun that barren shines,
Whose virtue on itself works no effect,
But in the fruitful Earth; there first receiv'd,
His beams, unactive else, their vigor find.
Yet not to Earth are those bright Luminaries
Officious, but to thee, Earth's habitant.
And, for the Heav'n's wide Circuit, let it speak
The Maker's high magnificence, who built
So spacious, and his Line stretched out so far;
That Man may know he dwells not in his own;
An Edifice too large for him to fill,
Lodg'd in a small partition, and the rest
Ordain'd for uses to his Lord best known.
The swiftness of those Circles attribute,
Though numberless, to his Omnipotence,
That to corporeal substances could add
Speed almost Spiritual; mee thou think'st not slow,
Who since the Morning hour set out from Heav'n
Where God resides, and ere mid-day arriv'd
In *Eden*, distances inexpressible
By Numbers that have name. But this I urge,
Admitting Motion in the Heav'ns, to show
Invalid that which thee to doubt it mov'd;
Not that I so affirm, though so it seem
To thee who hast thy dwelling here on Earth.
God to remove his ways from human sense,
Plac'd Heav'n from Earth so far, that earthly sight,
If it presume, might err in things too high,
And no advantage gain. What if the Sun
Be Centre to the World, and other stars,

By his attractive virtue and their own
Incited, dance about him various rounds?
Their wandering course, now high, now low, then hid.
Progressive, retrograde, or standing still,
In six thou seest; and what if, sev'nth to these,
The Planet Earth, so steadfast though she seem,
Insensibly three different Motions move?
Which else to several Spheres thou must ascribe,
Mov'd contrary with thwart obliquities,
Or save the Sun his labor, and that swift
Nocturnal and Diurnal rhomb suppos'd,
Invisible else above all Stars, the Wheel
Of Day and Night; which needs not thy belief,
If Earth industrious of herself fetch Day
Travelling East, and with her part averse
From the Sun's beam meet Night, her other part
Still luminous by his ray. What if that light,
Sent from her through the wide transpicuous air,
To the terrestrial Moon be as a Star,
Enlight'ning her by Day, as she by Night
This Earth? reciprocal, if land be there,
Fields and Inhabitants: Her spots thou seest
As Clouds, and Clouds may rain, and Rain produce
Fruits in her soft'n'd Soil, for some to eat
Allotted there; and other Suns perhaps
With their attendant Moons thou wilt descry
Communicating Male and Female Light,
Which two great Sexes animate the World,
Stor'd in each Orb perhaps with some that live.
For such vast room in Nature unpossessed
By living Soul, desert and desolate,
Only to shine, yet scarce to contribute
Each Orb a glimpse of Light, convey'd so far
Down to this habitable, which returns
Light back to them, is obvious to dispute.
But whether thus these things, or whether not,
Whether the Sun predominant in Heav'n
Rise on the Earth, or Earth rise on the Sun,
Hee from the East his flaming road begin,

Or Shee from West her silent course advance
With inoffensive pace that spinning sleeps
On her soft Axle, while she paces Ev'n,
And bears thee soft with the smooth Air along,
Solicit not thy thoughts with matters hid,
Leave them to God above, him serve and fear.   *[VIII, 66–168]*

HENRY MORE
(1614–87)

From *The Argument of Democritus Platonissans, or The Infinity of Worlds*

21

I will not say our world is infinite,
But that infinity of worlds there be;
The Centre of our world's the lively light
Of the warm sunne, the visible Deity
Of this externall Temple. *Mercurie*
Next plac'd and warm'd more throughly by his rayes,
Right nimbly 'bout his golden head doth fly:
Then *Venus* nothing slow about him strayes,
And next our *Earth* though seeming sad full sprightly playes.

22

And after her *Mars* rangeth in a round
With fiery locks and angry flaming eye,
And next to him mild *Jupiter* is found,
But *Saturn* cold wons in our outmost sky.
The skirts of his large Kingdome surely ly
Near to the confines of some other worlds

Whose Centres are fixèd starres on high,
'Bout which as their own proper Suns are hurld
Joves, Earths, and *Saturns:* round on their own axes twurld.

23

Little or nothing are those starres to us
Which in the azure Evening gay appear
(I mean for influence) but judicious
Nature and carefull Providence her dear
And matchlesse work did so contrive whileere,
That th' Hearts or Centres in the wide world pight
Should such a distance each to the other bear,
That the dull Planets with collated light
By neighbour suns might chearèd be in dampish night.

24

And as the Planets in our world (of which
The sun's the heart and kernal) do receive
Their nightly light from suns that do enrich
Their sable mantle with bright gemmes, and give
A goodly splendour, and sad men relieve
With their fair twinkling rayes, so our worlds sunne
Becomes a starre elsewhere, and doth derive
Joynt light with others, cheareth all that won
In those dim duskish Orbs round other suns that run.

25

This is the parergon of each noble fire
Of neighbour worlds to be the nightly starre,
But their main work is vitall heat t' inspire
Into the frigid spheres that 'bout them fare;
Which of themselves quite dead and barren are,
But by the wakening warmth of kindly dayes,
And the sweet dewie nights, they well declare
Their seminall virtue, in due courses raise
Long hidden shapes and life, to their great Makers praise.

26

These with their suns I severall worlds do call,
Whereof the number I deem infinite:
Else infinite darknesse were in this great Hall
Of th' endlesse Universe; For nothing finite
Could put that immense shadow into flight.
But if that infinite Suns we shall admit,
Then infinite worlds follow in reason right,
For every Sun with planets must be fit,
And have some mark for his farre-shining shafts to hit.

* * *

58

If starres be merely starres, not centrall lights,
Why swell they into so huge bignesses?
For many (as Astronomers do write)
Our sun in bignesse many times surpasse.
If both their number and their bulks were lesse
Yet lower placèd, light and influence
Would flow as powerfully, & the bosome presse
Of the impregnèd Earth, that fruit from hence
As fully would arise, and lordly affluence.

59

Wherefore these fixèd Fires mainly attend
Their proper charge in their own Universe,
And onely by the by of court'sie lend
Light to our world, as our world doth reverse
His thankfull rayes so far as he can pierce
Back unto other worlds. But farre aboven,
Further than furthest thought of man can traverse,
Still are new worlds aboven and still aboven,
In th' endlesse hollow Heaven, and each world hath his Sun.

* * *

86

For when as once these starres are come so nigh
As to seem one, the Comet must appear

In biggest show, because more loose they lie
Somewhat spread out, but as they draw more near
The compasse of his head away must wear,
Till he be brought to his least magnitude;
And then they passing crosse he doth repair
Himself, and still from his last losse renew'd
Grows, till he reach the measure which we first had view'd.

87

And then farre-distanc'd they bid quite adiew,
Each holding on in solitude his way.
Ne any footsteps in the empty Blew
Is to be found of that farre-shining ray.
Which processe sith no man did yet bewray,
It seems unlikely that the Comets be
Synods of starres that in wide Heaven stray:
Their smallnesse eke and numerositie
Encreaseth doubt and lessens probabilitie.

88

A cluster of them makes not half a Moon,
What should such tennis-balls do in the skie?
And few'll not figure out the fashion
Of those round firie Meteors on high.
Ne ought their beards much move us, that do lie
Ever cast forward from the Morning sunne
Nor back-cast tayls turn'd to our Evening-eye,
That fair appear whenas the day is done:
This matter may lie hid in the starres shadowed Cone.

89

For in these Planets conflagration,
Although the smoke mount up exactly round,
Yet by the suns irradiation
Made thin and subtil no where else its found
By sight, save in the dim and duskish bound
Of the projected Pyramid opake;
Opake with darknesse, smoke and mists unsound

Yet gilded like a foggie cloud doth make
Reflexion of fair light that doth our senses take.

90

This is the reason of that constant site
Of Comets tayls and beards: and that there show's
Not pure Pyramidall, nor their ends seem streight
But bow'd like brooms, is from the winds that blow,
I mean Ethereall winds, such as below,
Men finden under th' Equinoctiall line.
Their widend beards this aire so broad doth strow
Incurvate, and or more or lesse decline:
If not let sharper wits more subtly here divine.

SAMUEL BUTLER

(1612–80)

From "Hudibras"

[A joke perpetrated on Sidrophel, the astrologer,
and Whachum, his assistant]

Those two together long had liv'd,
In Mansion prudently contriv'd;
Where neither Tree, nor House could bar
The free detection of a Star;
And nigh an Antient Obelisk
Was rais'd by him, found out by Fisk,[9]
On which was written, not in words,
But Hieroglyphick Mute of Birds,
Many rare pithy Saws concerning
The worth of Astrologick Learning:

[9] An alchemist of Ben Jonson's time.

From top of this there hung a rope,
To which he fastned Telescope;
The spectacles with which the Stars
He reads in smallest Characters.
It hapned as a Boy, one night,
Did fly his Tarsel of a Kite,
The strangest long-wing'd Hawk that flies,
That like a Bird of Paradise,
Or Heraulds Martlet, has no legs.[10]
Nor hatches young ones, nor lays Eggs;
His Train was six yards long, milk-white,
At th'end of which, there hung a Light,
Enclos'd in Lanthorn made of Paper,
That far off like a Star did appear.
This, Sidrophel by chance espi'd,
And with Amazement staring wide,
Bless us, quoth he! What dreadful wonder
Is that, appears in heaven yonder?
A Comet, and without a Beard?
Or Star, that ne'r before appear'd?
I'm certain, 'tis not in the Scrowl,
Of all those Beasts, and Fish, and Fowl,
With which, like Indian Plantations,
The Learned stock the Constellations;
Nor those that drawn for Signs have bin,
To th'Houses, where the Planets Inn.
It must be supernaturall,
Unless it be that Cannon-Ball,
That, shot in th'aire, point-blank, upright,
Was borne to that prodigious height,
That learn'd Philosophers maintain,
It ne'r came backwards, down again;
But in the Aery region yet,
Hangs like the Body of Mahomet.
For if it be above the Shade,
That by the Earths round bulk is made,

[10]Because bird of paradise specimens had been shipped to Europe with their legs removed, it was thought that they were in perpetual flight, and hence supernatural, which gave rise to their name.

'Tis probable, it may from far,
Appear no Bullet but a Star.

This said, He to his Engine flew,
Plac'd near at hand, in open view,
And rais'd it, till it level'd right.
Against the Glow-worm Tayl of Kite
Then peeping through, Bless us! (quoth he)
It is a Planet now I see;
And if I err not, by his proper
Figure, that's like Tobacco-Stopper,
It should be Saturn; yes, 'tis clear,
'Tis Saturn, But what makes he there?
He's got between the Dragons Tayl,
And further leg behind, 'oth' Whale;
Pray Heaven, divert the fatal Omen,
For 'tis a Prodigie not common,
And can no less than the World's end,
Or Natures funeral portend.
With that, He fell again to pry,
Through Perspective, more wistfully,
When by mischance, the fatal string
That kept the Tow'ring Fowl on wing,
Breaking, down fell the Star: Well shot,
Quoth Whachum, who right wisely thought
H'had level'd at a Star, and hit it:
But Sidrophel more subtle-witted,
Cry'd out, What horrible and fearful
Portent is this, to see a Star fall;
It threatens Nature, and the doom
Will not be long, before it come.   *[II, iii. 399–474]*

WALT WHITMAN
(1819–92)

When I Heard the Learn'd Astronomer

When I heard the learn'd astronomer,
When the proofs, the figures, were ranged in columns before
    me,
When I was shown the charts and diagrams, to add, divide,
    and measure them,
When I sitting heard the astonomer where he lectured with
    much applause in the lecture-room,
How soon unaccountable I became tired and sick,
Till rising and gliding out I wander'd off by myself,
In the mystical moist night-air, and from time to time,
Look'd up in perfect silence at the stars.

ROBERT BROWNING
(1812–89)

My Star

All that I know
    Of a certain star
Is, it can throw
    (Like the angled spar)
Now a dart of red,
    Now a dart of blue;

Till my friends have said
They would fain see, too,
My star that dartles the red and the blue!
Then it stops like a bird; like a flower, hangs furled:
They must solace themselves with the Saturn above it.
What matter to me if their star is a world?
Mine has opened its soul to me; therefore I love it.

GEORGE MEREDITH

(1828–1909)

Meditation under Stars

What links are ours with orbs that are
So resolutely far?—
The solitary asks, and they
Give radiance as from a shield:
Still at the death of day,
The seen, the unrevealed.
Implacable they shine
To us who would of Life obtain
An answer for the life we strain,
To nourish with one sign.
Nor can imagination throw
The penetrative shaft: we pass
The breath of thought, who would divine
If haply they may grow
As Earth; have our desire to know;
If life comes there to grain from grass,
And flowers like ours of toil and pain;
Has passion to beat bar,

Win space from cleaving brain;
The mystic link attain,
Whereby star holds on star.

Those visible immortals beam
  Allurement to the dream:
Ireful at human hungers brook
  No question in the look.
  Forever virgin to our sense,
  Remote they wane to gaze intense:
Prolong it, and in ruthlessness they smite
The beating heart behind the ball of sight:
  Till we conceive their heavens hoar,
  Those lights they raise but sparkles frore,
And Earth, our blood-warm Earth, a shuddering prey
To that frigidity of brainless ray.
Yet space is given for breath of thought
Beyond our bounds when musing: more
When to that musing love is brought,
And love is asked of love's wherefore.
'Tis Earth's, her gift; else have we naught:
Her gift, her secret, here our tie.
And not with her and yonder sky?
Bethink you: were it Earth alone
Breeds love, would not her region be
  The sole delight and throne
  Of generous Deity?

  To deeper than this ball of sight
Appeal the lustrous people of the night.
Fronting yon shoreless, sown with fiery sails,
  It is our ravenous that quails,
Flesh by its craven thirsts and fears distraught.
  The spirit leaps alight,
  Doubts not in them is he,
The binder of his sheaves, the same, the right:
Of magnitude to magnitude is wrought,
To feel it large of the great life they hold:
In them to come, or vaster intervolved,

The issues known in us, our unsolved solved:
That there with toil Life climbs the selfsame Tree,
Whose roots enrichment have from ripeness dropped.

So may we read and little find them cold:
Let it but be the lord of Mind to guide
Our eyes; no branch of Reason's growing lopped;
Nor dreaming on a dream; but fortified
By day to penetrate black midnight; see,
Hear, feel, outside the senses; even that we,
The specks of dust upon a mound of mold,
We who reflect those rays, though low our place,
    To them are lastingly allied.

So may we read, and little find them cold:
Not frosty lamps illumining dead space,
Not distant aliens, not senseless Powers.
The fire is in them whereof we are born;
The music of their motion may be ours.
Spirit shall deem them beckoning Earth and voiced
Sisterly to her, in her beams rejoiced.
Of love, the grand impulsion, we behold
    The love that lends her grace
    Among the starry fold.
Then at new flood of customary morn,
    Look at her through her showers,
    Her mists, her streaming gold,
A wonder edges the familiar face:
She wears no more that robe of printed hours;
Half strange seems Earth, and sweeter than her flowers.

EMILY DICKINSON

(1830–86)

It Troubled Me as Once I Was

It troubled me as once I was,
For I was once a child,
Deciding how an atom fell
And yet the heavens held.

The heavens weighed the most by far,
Yet blue and solid stood
Without a bolt that I could prove;
Would giants understand?

Life set me larger problems,
Some I shall keep to solve
Till algebra is easier
Or simpler proved above.

Then too be comprehended
What sorer puzzled me,
Why heaven did not break away
And tumble blue on me.

ROBINSON JEFFERS

(1887–1962)

Nova

That Nova was a moderate star like our good sun; it stored no
    doubt a little more than it spent
Of heat and energy until the increasing tension came to the
    trigger-point
Of a new chemistry; then what was already flaming found a
    new manner of flaming ten-thousandfold
More brightly for a brief time; what was a pin-point fleck on a
    sensitive plate at the great telescope's
Eye-piece now shouts down the steep night to the naked eye,
    a nine-day super-star.
                          It is likely our moderate
Father the sun will some time put off his nature for a similar
    glory. The earth would share it; these tall
Green trees would become a moment's torches and vanish,
    the oceans would explode into invisible steam,
The ships and the great whales fall through them like flaming
    meteors into the emptied abysm, the six mile
Hollows of the Pacific sea-bed might smoke for a moment.
    Then the earth would be like the pale proud moon,
Nothing but vitrified sand and rock would be left on earth.
    This is a probable death passion
For the sun's planets; we have no knowledge to assure us it
    may not happen at any moment of time.

Meanwhile the sun shines wisely and warm, trees flutter
    green in the wind, girls take their clothes off

To bathe in the cold ocean or to hunt love; they stand
    laughing in the white foam, they have beautiful
Shoulders and thighs, they are beautiful animals, all life is
    beautiful. We cannot be sure of life for one moment;
We can, by force and self-discipline, by many refusals and a
    few assertions, in the teeth of fortune assure ourselves
Freedom and integrity in life or integrity in death. And we
    know that the enormous invulnerable beauty of things
Is the face of God, to live gladly in its presence, and die
    without grief or fear knowing it survives us.

From *Margrave*

On the small marble-paved platform
On the turret on the head of the tower,
Watching the night deepen,
I feel the rock-edge of the continent
Reel eastward with me below the broad stars.
I lean on the broad worn stones of the parapet top
And the stones and my hands that touch them reel eastward.
The inland mountains go down and new lights
Glow over the sinking east rim of the earth.
The dark ocean comes up,
And reddens the western stars with its fog-breath
And hides them with its mounded darkness.

The earth was the world and man was its measure, but our
    minds have looked
Through the little mock-dome of heaven the telescope-slotted
    observatory eyeball, there space and multitude came in
And the earth is a particle of dust by a sand-grain sun, lost in
    a nameless cove of the shores of a continent.
Galaxy on galaxy, innumerable swirls of innumerable stars,
    endured as it were forever and humanity
Came into being, its two or three million years are a moment,
    in a moment it will certainly cease out from being

And galaxy on galaxy endure after that as it were forever . . .
    But man is conscious,
He brings the world to focus in a feeling brain,
In a net of nerves catches the splendor of things,
Breaks the somnambulism of nature . . . His distinction
    perhaps,
Hardly his advantage. To slaver for contemptible pleasures
And scream with pain, are hardly an advantage.
Consciousness? The learned astronomer
Analyzing the light of most remote star-swirls
Has found them—or a trick of distance deludes his prism—
All at incredible speeds fleeing outward from ours.
I thought, no doubt they are fleeing the contagion
Of consciousness that infects this corner of space.

          *   *   *

For often I have heard the hard rocks I handled
Groan, because lichen and time and water dissolve them,
And they have to travel down the strange falling scale
Of soil and plants and the flesh of beasts to become
The bodies of men; they murmur at their fate
In the hollows of windless nights, they'd rather be anything
Than human flesh played on by pain and joy,
They pray for annihilation sooner, but annihilation's
Not in the book yet.
          So, I thought, the rumor
Of human consciousness has gone abroad in the world,
The sane uninfected far-outer universes
Flee it in a panic of escape, as men flee the plague
Taking a city.
          You would be wise, you far stars,
To flee with the speed of light this infection.
For here the good sane invulnerable material
And nature of things more and more grows alive and cries.
The rock and water grow human, the bitter weed
Of consciousness catches the sun, it clings to the near stars,
Even the nearer portion of the universal God
Seems to become conscious, yearns and rejoices
And suffers: I believe this hurt will be healed

Some age of time after mankind has died,
Then the sun will say "What ailed me a moment?" and
    resume
The old soulless triumph, and the iron and stone earth
With confident inorganic glory obliterate
Her ruins and fossils, like that incredible unfading red rose
Of desert in Arizona glowing life to scorn,
And grind the chalky emptied seed-shells of consciousness
The bare skulls of the dead to powder, after some million
Courses around the sun her sadness may pass:
But why should you worlds of the virgin distance
Endure to survive what it were better to escape?

*   *   *

On the little stone-girdled platform
Over the earth and the ocean
I seem to have stood a long time and watched the stars pass.
They also shall perish I believe.
Here to-day, gone to-morrow, desperate wee galaxies
Scattering themselves and shining their substance away
Like a passionate thought. It is very well ordered.

ROBERT FROST
(1875–1963)

The Star-Splitter

"You know Orion always comes up sideways.
Throwing a leg up over our fence of mountains
And rising on his hands, he looks in on me
Busy outdoors by lantern-light with something
I should have done by daylight, and indeed,

After the ground is frozen, I should have done
Before it froze, and a gust flings a handful
Of waste leaves at my smoky lantern chimney
To make fun of my way of doing things,
Or else fun of Orion's having caught me.
Has a man, I should like to ask, no rights
These forces are obliged to pay respect to?"
So Brad McLaughlin mingled reckless talk
Of heavenly stars with hugger-mugger farming,
Till having failed at hugger-mugger farming,
He burned his house down for the fire insurance
And spent the proceeds on a telescope
To satisfy a life-long curiosity
About our place among the infinities.

"What do you want with one of those blame things?"
I asked him well beforehand. "Don't you get one!"
'Don't call it blamed; there isn't anything
More blameless in the sense of being less
A weapon in our human fight,' he said.
'I'll have one if I sell my farm to buy it."
There where he moved the rocks to plough the ground
And ploughed between the rocks he couldn't move,
Few farms changed hands; so rather than spend years
Trying to sell his farm and then not selling,
He burned his house down for the fire insurance
And bought the telescope with what it came to.
He had been heard to say by several:
'The best thing that we're put here for's to see;
The strongest thing that's given us to see with's
A telescope. Someone in every town
Seems to me owes it to the town to keep one.
In Littleton it may as well be me.'
After such loose talk it was no surprise
When he did what he did and burned his house down.

Mean laughter went about town that day
To let him know we weren't the least imposed on,
And he could wait—we'd see to him tomorrow.

But the first thing next morning we reflected
If one by one we counted people out
For the least sin, it wouldn't take us long
To get so we had no one left to live with.
For to be social is to be forgiving.
Our thief, the one who does our stealing from us,
We don't cut off from coming to church suppers,
But what we miss we go to him and ask for.
He promptly gives it back, that is if still
Uneaten, unworn out, or undisposed of.
It wouldn't do to be too hard on Brad
About his telescope. Beyond the age
Of being given one for a Christmas gift,
He had to take the best way he knew how
To find himself in one. Well, all we said was
He took a strange thing to be roguish over.
Some sympathy was wasted on the house,
A good old-timer dating back along;
But a house isn't sentient; the house
Didn't feel anything. And if it did,
Why not regard it as a sacrifice,
And an old-fashioned sacrifice by fire,
Instead of a new-fashioned one at auction?
Out of a house and so out of a farm
At one stroke (of a match), Brad had to turn
To earn a living on the Concord railroad,
As under-ticket-agent at a station
Where his job, when he wasn't selling tickets,
Was setting out up track and down, not plants
As on a farm, but planets, evening stars
That varied in their hue from red to green.

He got a good glass for six hundred dollars.
His new job gave him leisure for star-gazing.
Often he bid me come and have a look
Up the brass barrel, velvet black inside,
At a star quaking in the other end.
I recollect a night of broken clouds
And underfoot snow melted down to ice,

And melting further in the wind to mud.
Bradford and I had out the telescope.
We spread our two legs as we spread its three,
Pointed our thoughts the way we pointed it,
And standing at our leisure till the day broke,
Said some of the best things we ever said.
That telescope was christened the Star-splitter,
Because it didn't do a thing but split
A star in two or three the way you split
A globule of quicksilver in your hand
With one stroke of your finger in the middle.

It's a star-splitter if there ever was one
And ought to do some good if splitting stars
'Sa thing to be compared with splitting wood.

We've looked and looked, but after all where are we?
Do we know any better where we are,
And how it stands between the night tonight
And a man with a smoky lantern chimney?
How different from the way it ever stood?

W. H. AUDEN

(1907–73)

Do We Want to Return to the Womb?

Do we want to return to the womb? Not at all.
No one really desires the impossible:
That is only the image out of our past
We practical people use when we cast
Our eyes on the future, to whom freedom is

The absence of all dualities.
Since there never can be much of that for us
In the universe of Copernicus,
Any heaven we think it decent to enter
Must be Ptolemaic with ourselves at the centre.

MURIEL RUKEYSER
(1913—80)

Brecht's Galileo

Brecht saying: Galileo talking astronomy
Stripped to the torso, the intellectual life
Pouring from the gross man in his nakedness.

Galileo, his physical contentment
Is having his back rubbed by his student: the boy mauls:
The man sighs and transforms it: intellectual product!

Galileo spins a toy of the earth around
The spinning sun: he looks at the student boy.
Learning is teaching, teaching is learning.
Galileo
Demonstrates how horrible is betrayal.
Particularly on the shore of a new era.

ADRIENNE RICH

(1929—)

Planetarium

Thinking of Caroline Herschel (1750–1848)
astronomer, sister of William, and others.

A woman in the shape of a monster
a monster in the shape of a woman
the skies are full of them

a woman   'in the snow
among the Clocks and instruments
or measuring the ground with poles'
in her 98 years to discover
8 comets

she whom the moon ruled
like us
levitating into the night sky
riding the polished lenses

Galaxies of women, there
doing penance for impetuousness
ribs chilled
in those spaces   of the mind

An eye,

       'virile, precise and absolutely certain'
       from the mad webs of Uranusborg[11]

[11] Uranienborg was Tycho Brahe's observatory and the NOVA was the new star in
Cassiopeia he discovered in 1573.

encountering the NOVA
every impulse of light exploding
from the core
as life flies out of us

Tycho whispering at last
'Let me not seem to have lived in vain'

What we see, we see
and seeing is changing

the light that shrivels a mountain
and leaves a man alive

Heartbeat of the pulsar
heart sweating through my body

The radio impulse
pouring in from Taurus

I am bombarded yet   I stand

I have been standing all my life in the
direct path of a battery of signals
the most accurately transmitted most
untranslatable language in the universe
I am a galactic cloud so deep   so invo-
luted that a light wave could take 15
years to travel through me   And has
taken   I am an instrument in the shape
of a woman trying to translate pulsations
into images   for the relief of the body
and the reconstruction of the mind.

MAY SWENSON
(1919—)

Science and Religion—A Merger

When Galileo Galilei first turned a telescope on the heavens,
                        Was St. Peter buried on Vatican hill,
400 years ago, his revelations were astounding. Jupiter
                   the site of the great Roman Catholic basilica
he found, has its own miniature system of planets,
                that bears his name? Last week Pope Paul. . . .
or moons. He saw the mountains of the moon,
            gave his support to that theory, announcing that bones
spots on the sun and the crescent shape of Venus.
                discovered in 1953 under the basilica
He found that the Milky Way Galaxy
           had been identified to his satisfaction as those of
of which we are a part is actually formed from billions of
          the saint. For Christians. . . . it is not an idle question . . .
distant, dim stars. Since then, telescopes have gradually increased
            The claims. . . . rest on two arguments
in size and quality, culminating in 1948
           concerning Peter: First, that the statement
in completion of the great reflector on Mount Palomar
         of Jesus quoted by Matthew: "Thou art Peter, and
in California. This instrument, with a parabolic mirror
          upon this rock I shall build my church"
200 inches in diameter, has been to modern astronomy what
         is literally true . . . and second, that the apostle
Galileo's instrument was to science in the 17th Century.
         Peter was bishop of Rome, and thus the first

It has carried man's ken toward the outer fringes

in an unending series of Roman bishops—or—popes,

of the universe and it has enlarged his knowledge of the galaxies.

who embody the full authority to guide

It first identified the strange quasars that seem to be the

the Christian Church. In 1939, the

most distant observable objects and the light-collecting power of

Vatican excavations beneath the main altar of St. Peter's

its huge mirror has brought into view peculiar stars,

began to uncover a series of tombs, which

that, while not very distant, are too dim to be observed with

was held to include the tomb of Peter. But

other instruments. . . . While others are being built, none comes close to

the first announcement

the 200-inch Hale Telescope—with one exception. That is

came only in 1949, when Pope Pius XII stated that

the 236-inch reflector being built by the Soviet Union near Zelenchuk

an urn containing the remains of the apostle

in the Caucasus. . . . Apparently the Russians hope to dazzle the world

had been uncovered. . . . Later, however, the bones

as they did with their Sputnik in 1957, by a surprise announcement after

in the urn were shown to be that of a woman.

their first look into the realms previously beyond reach. . . .

During the 1950's, Professor Margherita Guarducci,

While the Russians, with their new instrument, will be able to see things

a Vatican expert on inscriptions, argued that

no one else can, their field of view will be limited by

writings on walls beneath the altar pointed to

their geography. . . . Because almost all of the world's great observatories

a particular niche as the resting place

are north of the Equator, the southern part of the sky

of Peter's remains. Earlier a team of Vatican archeologists

is by far the least explored. The center

had reported secretly to the Pope that the niche,

of the Milky Way Galaxy lies there

and a box in it, were empty. But Professor Guarducci—

plus the two nearest baby-galaxies (the Clouds

persisted, reporting that Monsignor Kaas, then secretary

of Magellan). . . . One of the dreams of American astronomers and administrator for the Fabric of St. Peter's, told her

is the placing of a large telescope

that he and two workmen had removed some bones

into orbit above the earth's atmosphere.

from the niche without the knowledge of Vatican

This has become possible with the giant Saturn

archeologists. It was these bones that Pope Paul

rockets designed to send men to

last week identified as those of St. Peter.

the moon. Our present view of

European archeologists familiar with the

the heavens can be likened to that of a lobster beneath the

Vatican diggings remain privately skeptical but publicly

murky waters of Long Island Sound.

silent. . . . Further investigations will likely be

A telescope above the ocean of air would open

colored by the Pope's decision to commit some of

new realms of knowledge concerning our

his prestige to a circumstantial argument the

nearest neighbors in space, as well as

bones in question are indeed Peter's. The Vatican

the nature of the universe as a whole.

has got itself into a position where its case can't be

However, as with other grandiose science projects, the problem is

proved scientifically, an American archeologist . . .

cost. . . . American action may be delayed until

said last week. He said, "We'll probably never know

the Russians have done it first.

whose bones they are."

DORIS LESSING

(1919—)

From *Briefing for a Descent into Hell*

As I thought that I would like to see the earth speed up a little, but not as fast as before, when a year's turn around the sun seemed like the spin of a coin, it did speed up—and now I saw other patterns of light, or colour, deepen and fade and marry and merge and move, and as I thought that all these patterns were no more than a composite of the slower individual pulses and currents I had seen earlier, and that they were making up the glowing coloured mist that was the envelope of the globe, it came into my mind that the glowing envelope of the globe seemed to be set, or held, by something else, just as it, in its place, held the rhythms of the earth, our earth. My mind made another outwards-going, outswelling, towards comprehension, and now I saw how lines and currents of force and sympathy and antagonism danced in a web that was the system of planets around the sun, so much a part of the sun that its glow of substance, lying all about it in space, held the planets as intimately as if these planets were merely crystallizations or hardening of its vaporous stuff, moments of density in the solar wind. And this web was an iron, a frightful necessity, imposing its design.

Now I watched, as the earth turned fast, but still so that I could see the change and growth and dying away of patterns, how as the planets moved and meshed and altered and came closer to each other and went away again, exerting a pressure of forces on each other that bound them all, on the earth the little crusts of matter that were men, that were humanity, changed and moved. Just as the waters, the oceans (a little film of fluid matter on the big globe's surface) moved and swung under the compulsion of the sun and the moon, so did the life of man, oscillating in its web of necessity, in its place in the life of the planets, a minute crust on the surface of a thickening and becoming visible of the Sun's breath that was called Earth. Humanity was a pulse in the life of the Sun, which lay burning there in a vast white explosion of varying kinds of light, or sound, some stronger and thicker, some tenuous, but at all forces and strengths, which fluid lapped out into space holding all these crumbs and drops and little flames in a dance—and the force that held them there, circling and whirling in their dance, was the Sun, the energy of the Sun, and that was the controlling governor of them all, beside whose strength, all the subsidiary laws and necessities were nothing. The ground and soul and heart and centre of this little solar system was the light and pulse and song of the Sun, the Sun was King. But although this central strength, this magestic core of our web, was an essence to the whole system, further out and away from the centre, where poor dark Pluto moved, perhaps it might be that the tug and pull and pressure of the planets seemed more immediate; perhaps out there, or further, the knowledge that the Sun is still the deep low organ note that underlies all being is forgotten—forgotten more even than on earth, spinning there so crooked and sorrowful and calamitous with its weight of cold necessity so close. And perhaps, or so I thought as I saw the dance of the sun and its attendants, Mercury the Sun's closest associate was the only one which could maintain steadily and always the consciousness of the sun's underlying song, its need, its intention, Mercury whose name was, also, Thoth, and Enoch, Buddha, Idris, and Hermes, and many other styles or titles in the earth's histories, Mercury the Messenger, the carrier of news, or information from the Sun, the disseminator of laws from God's singing centre.

Yes, but farther out, on the third crookedly spinning planet, it is harder to keep that knowledge, the sanity and simplicity of the great Sun, and indeed poor Earth is far from grace, and so it was easy to see, for at that tempo of spin that enabled me to watch clearly the marrying of events on earth and in the rest of its fellow planets, I watched how wars and famines, and earthquakes and disasters, floods and terrors, epidemics and plagues of insects and rats and flying things came and went according to the pressures from the combinations of the planets and the sun—and the moon. For a swarm of locusts, a spreading of viruses, like the life of humanity, is governed elsewhere. The life of man, that little crust of matter, which was not even visible until one swooped down close as a bird might sweep in and out for a quick survey of a glittering shoal of fish that puckered a wave's broad flank, that pulse's intensity and size and health was set by Mercury and Venus, Mars and Jupiter, Saturn, Neptune, Uranus, and Pluto, and their movements, and the Centre of light that fed them all.

PHYSICS

INTRODUCTION

I believe that whatever we do or live for has its
causality; it is good, however, that we cannot see
through to it.

—Albert Einstein

The earliest physics dealt with the material world in general
and was equated with natural science. This equation can be
seen in Aristotle's view of why stones fall to the ground: They
tend to go where they belong. Physics as the science of matter, energy, and the interaction of the two, really began in the
Renaissance. The unprecedented construction of divine cathe-
drals—and their attendant building disasters—pointed out
the need for knowledge of worldly laws while practical engi-
neering concerns kindled an interest in basic physical laws.
Desires for advanced weaponry added impetus to this search
for knowledge. In short, it was out of fundamental engineer-
ing and technological needs that physics as an exact science
developed.

The one thing that held back physics, however, was the
lack of an adequate system of mathematics. Once isolated,
problems could not be solved until the mathematical knowledge had been achieved. As the math was worked out, however, physics accelerated at a rate that could not have been
predicted. This great thrust began in Isaac Newton's time
(1642–1727), when he and other mathematicians such as Gott-
fried Leibnitz produced the necessary tools of calculus.

Two of the topics that occupied the attention of physicists
in the Newtonian era were gravity and optics. Studies of these
phenomena in turn intrigued the poets, for whom gravity was

associated with seriousness and light with spirituality. Gravity deals with the attractive forces that bodies exert on one another due to their mass. Another attractive force that intrigued even the ancient writers (who had no concept of gravity) was the magnet.

In the earliest days of physics as a natural science, Pliny the Elder described a magnet's attractive property as "mysterious and unseen." While he noted the remarkable property of the magnet, its connection with the heavens and polarity went unnoticed. Over a thousand years after Pliny, a Gallic scholar named Pierre de Maricourt, also called Petrus Peregrinus (Peter the Pilgrim) because he once went to the Holy Land, again turned attention toward the magnet. He was an experimentalist who believed that he had designed a perpetual motion machine based on the attractive powers of the magnet. In 1269 he wrote a long letter describing this machine to a friend who was unschooled in science. So that his friend would understand the machine, this medieval scientist described the properties of the magnet in detail. Had his friend known more about the magnet, we would not have had his remarkable treatise on the subject.

Among Petrus Peregrinus' important observations is a description of how to tell the north pole of a magnet from the south; he was also the first to place a magnetized needle on a pivot so that it could be used as a compass. Throughout his letter he is remarkably free of superstition concerning the magnet. Those who followed him would not be so empirical.

When Thomas Browne, English author and physician, wrote on the lodestone in *Pseudodoxia Epidemica* (1646), he clearly recognized the use of the lodestone for navigation and astronomy but discussed polarity as a creation of God intended to give the world a permanent orientation. The metaphysical poet Thomas Stanley viewed the magnet as having a mysterious romantic power to attract and compel an obstinate lover who was quite literally as hard as nails. The Renaissance lyric had a strong inclination toward unusual metaphors, and one particular school of "metaphysical poetry" specialized in expanding its exotic metaphors into complex logical arguments.

It is no wonder that all of these poets, given to model building and logic, were immensely stimulated by current science. They were also religious men who harbored certain fears that science had the power to deflect man's intellect from the study of God. While George Herbert expressed this fear in his poem "Vanitie," Henry Vaughan made a conscious attempt to reconcile natural science and divine contemplation in his poem "The Starre." Vaughan's argument again hinges on magnetic attraction: The star is drawn to beauty and purity. "These are the Magnets which so strongly move./And work all night upon thy light and love." The star takes its commandments, its motion in the heaven, from God, and Vaughan offers the star as an example of man's duty to be magnetized by celestial desire and moved by God.

Francis Quarles, whose book *Emblems* (1635) was possibly the most popular collection of verse in the seventeenth century, was another poet to use the magnet as a religious metaphor. His exquisite description of the quivering compass point—the "arctic needle"—reminds him of the frantic soul moved at random until it has the guidance of God's true north. Like Vaughan, Quarles also draws a parallel between the moon enlightened by the sun and man enlightened by God, the "blessed loadstone."

Indeed, this unseen attractive force was viewed as wondrous by poets. But when Newton presented his theory of gravitation, there were some seventeenth-century scientists who rejected it as impossible because it involved action at a distance, which was magic!

Nevertheless, both the poets and the scientists continued to make allusions to the working of a lodestone and the cosmic forces of attraction that kept the heavenly spheres on course. Jonathan Swift, ever the satirist of empirical science, devised his own applications of magnetic theory. In Book III of *Gulliver's Travels* he describes a Flying Island of astronomers with a lodestone set into its adamant core. By means of the magnet the island rises and falls; Swift's adamant island is also politically hard, and the scientific metaphor intended to strike at the Royal Society is also directed against the English political rule of Ireland, which is depicted as a ground country held in check by the Flying Island. If all else fails, the final strategy is to drop the island itself and destroy the whole country.

Swift's allegory is faithful to current astronomy. In 1685 Edmund Halley had presented to the Royal Society "the Effects of Gravity on the Descent of Heavenly Bodies. . . ," making reference to an analogy with lodestones that was commonly connected with Newton's gravitational theory, to which he pays tribute.

Although Swift and his friend Alexander Pope generally scoffed at science, Pope's famous epigram on Newton neatly summarizes the awesome reverence Newton inspired for his age: "Nature and Nature's law lay hid in night: God said, Let Newton be! And all was light."

In addition to his work on gravitation, Newton made the first significant contribution to the study of light in over two thousand years. Scientists since Aristotle had known that when light went through glass it changed color, but their explanation had left something to be desired—light darkens a little at the thin end of the glass and becomes red; it is darkened more where the glass is thicker and becomes green, etc. This was not an explanation but a description of the problem. It failed to explain why the light came in as a disk (if the sun is the source) and left the glass as an oblong.

During the plague years of 1665–66 Newton did his thinking and experimenting in optics (the *Optics* was not published until 1704). From his work with prisms he realized that light was not modified but separated.

I procured me a Triangular glass-Prisme to try therewith the celebrated *Phaenomena of Colours*. And in order thereto having darkened my chamber, and made a small hole in my window-shuts, to let in a convenient quantity of the Sun's light, I placed my Prisme at his entrance, that it might be thereby refracted to the opposite wall. It was at first a very pleasing divertisement, to view the vivid and intense colours produced thereby; but after a while applying my self to consider them more circumspectly, I became surprised to see them in an *oblong* form; which, according to the received laws of Refraction, I expected should have been *circular*.

And I saw . . . that the light, tending to [one] end of the Image, did suffer a Refraction considerably greater then the light tending to the other. And so the true cause of the length of that Image was detected to be no other, then the *Light* consists of *Rays differently refrangible,* which, without any respect to a difference in their incidence, where, according to their degrees of refrangibility, transmitted towards divers parts of the wall.

Blue is refracted more than red, and this is an absolute property of the colors. Newton then made a crucial experiment. He took light that had gone through a prism and put it through a second prism. If the glass was crucial in adding color to white daylight, then it would modify the color of the light a second time, but if it was really a property of the light that causes the spectrum, then the color of the light would not be altered by the second prism. Newton discovered that separated colors cannot be changed further:

Then I placed another Prisme . . . so that the light . . . might pass through that also, and be again refracted before it arrived at the wall. This done, I took the first Prisme in my hand and turned it to and fro slowly about its *Axis,* so much as to make the several parts of the Image . . . successively pass through . . . that I might observe to what places on the wall the second Prisme would refract them.

When any one sort of Rays hath been parted from those of other kinds, it hath afterwards obstinately retained its colour, notwithstanding my utmost endeavours to change it.

Newton understood scientific method and knew how to test a theory. In a letter to the Royal Society in 1672 he wrote, "A naturalist would scarce expect to see ye science of those colours become mathematical, and yet I dare affirm that there is as much certainty in it as in any other part of Opticks."

Newton's study of color spread to the arts. Although religious poets of the seventeenth century—like Francis Quarles—were apt to view the prism as an emblem of vanity, Newton's prismatic light became a poetic subject. James Thomson, who thought of Newton as "Our philosophic Sun," saw the "watery" rainbow as a dramatic performance of revealed color. The vocabulary of color, simple words like *orange* and *violet*, were given by Newton a new value that linked poetry to scientific observation.

In the century that followed there was some controversy over this connection. While Vincent Bourne's epigram on "Newton's *Principia*" recognized mathematics as a key to opening the unknown, Alexander Pope, and later William Blake, regarded the rational and doubting mentality of the experimental scientist as a threat to pure belief. Religious poets generally disapproved of probing heavenly secrets, but William Cowper found a compromise in admiring Newton as a sage whose investigation of the shining light drew men closer to God.

While Newton made tremendous advances in the study of light, there was one basic limiting factor. He viewed light as being made up of particles that moved in straight lines. Here Newton had a problem he really could not explain. When light strikes a glass, some of it is reflected back into the air and some is transmitted or refracted through the glass. What determines whether a particle will bounce off or be transmitted? Newton had no satisfactory answer. He suggested that the particles worked in fits. Sometimes a particle would bounce off and sometimes under identical circumstances it would be transmitted. However, in the early nineteenth century, experiments involving interference and diffraction of light made it clear that light was a wave phenomenon.

The question now raised was "What was waving?" In the mid-1880s Scottish physicist James Clerk Maxwell (1831–79) theorized that the vibrations associated with light waves were due to electromagnetic fields. The notion of fields was foreign to the Newtonian world view. Newton viewed gravitational attraction as a pulling of the planets across space—a mysterious action at a distance. Maxwell's field theory required that there be a medium for conducting light, just as sound waves are conducted by the air molecules. This led to the postulation of an invisible jelly surrounding the earth that served as the medium for light conduction, the "ether."

In 1887 the famous Michelson-Morley experiment was carried out to prove the existence of the ether, which was a topic of considerable debate among physicists. The argument was based on the simple principle that it takes a longer time to row a boat a hundred yards across a stream than it does to row a boat a hundred yards downstream. If there were, in fact, an ether, then it would take a light wave a longer time to go a fixed distance across the ether's current, which would be caused by the passage of the earth through space, than to go the same distance with the ether's current. Or, if the light went against the current, then it would take longer to cover the specified distance. In the Michelson-Morley experiment two light waves were sent out simultaneously at right angles to each other, one going across the current, the other either with or against the current. At some point in the travels of the light waves they would be out of phase if there was, in fact, an ether. The results of the experiment failed to indicate an ether.

In 1905 Einstein published a paper that resolved the difficulties presented by the results of the Michelson-Morley experiment. This paper, in which Einstein presented his special theory of relativity, has changed man's view of space and time forever.

Einstein began the paper by discussing Michael Faraday's theory of induction and attempted to show why results obtained in experiments were not in accord with the theory. On the basis of Faraday's theory of induced currents, when a magnet is rotated about a conductor, an electric current should be induced between the magnet and the core, but not when the core is rotated within the magnet. This, however, was not supported by the empirical data. It didn't matter which was rotated, a current always resulted. Einstein pointed out that the current depended only on relative movement, not upon absolute movement. He went on to suggest that the phenomena of electromagnetism and mechanics indicate that there is no property that corresponds to the notion of absolute rest. He then joined the two realms of mechanics and electromagnetism by postulating that the speed of light is constant to all observers regardless of their relative motion. Several important consequences followed immediately. Since the special theory of relativity postulated that the speed of light was a constant, even if there were an ether drift the use of light would not be adequate to detect it. This explains the null results of the Michelson-Morley experiment.

If the velocity (v) is a constant and $v = distance \neq time$, then both time and space are the variables, or relative, not absolute. No distance or time can be meaningfully specified without reference to the observer. The second postulate presented in this paper is that the laws of physics are the same for all observers who are moving at constant speed, relative to each other. As a consequence of these two postulates Einstein deduced two revolutionary results. First, a clock in motion moves more slowly than a clock at rest. This is not a mechanical property of the clock, but one of time and space. Second, the length of a moving object is contracted relative to its length at rest. As an object reaches a velocity of nine-tenths the speed of light it has contracted to half its length at rest. It would shrink to nothing as it approaches the speed of light, which is why 186,000 miles per second is the limiting velocity of the physical world.

Another concept that is altered radically by the special theory of relativity is that of simultaneity. There can be no meaningful notion of simultaneity between two systems in relative motion. Consider a sailor on a moving ship. Let us assume that on the shore there are two lighthouses and midway between the lighthouses is an observer, O, who can turn the lights on in both towers simultaneously. Let us also assume that when the observer turns the lights on, the sailor is equidistant from both lighthouses (see opposite). When the switch is thrown and both lights go on, what will be seen by the observer and the sailor? To the observer, light from both sources will reach him at the same time and he will see that both lights went on simultaneously. To the sailor, it will appear somewhat different. If the lights *are* on at T_1, in the interval it takes the light to travel from the lighthouses to the ship, the ship will have changed its position and the light from the tower on the left will reach the sailor at T_3 while the light from the tower to the right will reach the sailor earlier at T_2. Thus, to the sailor, one light goes on before the other. This effect is appreciable only at great speeds.

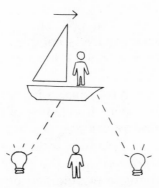

In the 1905 paper Einstein showed that two of the three basic concepts in physics, that is, length and time, were relative to motion. In a follow-up paper he discussed matter and its relation to motion. He explained that matter is merely congealed energy and that if a body gives off energy, E, its mass is dimished E/c^2. The mass of a body is a measure of its energy. This yields the famous $E = mc^2$ equation, where energy is in ergs, mass in grams, and the speed of light in centimeters per second.

Further, as the speed of a given body increases, its mass increases. The mass of a moving body is its mass at rest divided by $\sqrt{1-v^2/c^2}$. This links the conservation of energy with the conservation of matter. It also explains why electrons are more massive when moving than when at rest. It shows that mass is another form of energy, which has been strikingly verified by the nuclear devices of our time. Energy is merely liberated mass, and photons, or light quanta, are particles that have shed their mass and are traveling at the speed of light.

It is a speed to which we have rapidly adjusted in the brief time since relativity unleashed the universe from its hitching post. In 1952 Einstein expressed the fear that this age of "dull specialization that stares with self-conceit through horn-rimmed glasses . . . destroys poetry." Einstein, in fact, liked to write playful verses and often responded to occasions or provocative thoughts in rhymed doggerel. One such quatrain found among his papers can be taken as an epigram on subjectivity. It comes from the same wit that reduced his complex theory of relativity to the simple perception that "one man's space is another man's time."

That little word "WE" I mistrust, and here's why:
No man of another can say "He is I."
Behind all agreement lies something amiss
All seeming accord cloaks a lurking abyss.

The poem should have calmed his fears about this age kill-
ing poetry, for in essence it explains why the world view expressed in the new physics is conducive to poetry. It is a
world of paradox, and paradox invites metaphor. As we have
seen, even poets responding to the language of classical physics were able to derive a double and often symbolic meaning
from concepts such as *gravity*, with its implication of seriousness, and *magnetism*, with its implication of personal attraction. The language of particle physics expands the poetic
vocabulary with its *quarks, dimensions, relativity, masses, energy,
breakdowns, and half-lives.* It is a rich word-hoard that we can
readily and picturesquely apply to the human condition. If
particles *break down,* so do people, and if the decay of radioactive elements can be expressed in *half-lives,* so too—as an Erica
Jong poem suggests—can the decay of human wholeness be
expressed in the scientific language of *particular* destruction.
Jong writes in "Half-Life" as a metaphysical poet drawing out
a long conceit that compares making love to atomic war. She
is conscious of making the metaphor and conscious also that
the reality of the atomic bomb has intensified the connection
between love and individual death that poets have always understood and expressed metaphorically.

Einstein himself believed in the common pursuit of art and
science, though he attempted to distinguish between the intuitive language of art and the logic of science. Since Einstein,
the acceptance of uncertainty (which he rejected) as a principle of physics has altered the language of physics. As many
writers have pointed out, the nature of its investigation has
rendered the language of physics metaphysical and poetic.

Like the metaphysical poets of the seventeenth century,
contemporary poets have adopted not only the vocabulary of
science but also its instruments as metaphors for probing.
Muriel Rukeyser, for example, uses the gyroscope, with its
rotating wheel mounted such that its axis can turn freely in all

directions, as a concept to explain "the dynamics of desire." Its movement is her orbit of thought; its motion is the motion of a rooted mind in an expanding universe. The paradox of its torque leads her to the tension field of man rooted to earth but ready to fly.

Rukeyser's conclusion is reminiscent of the German poet Rainer Maria Rilke, who once spoke of "a tune beyond us, yet ourselves." In a sense the new physics is that tune: mysterious to all but its specialized investigators, yet alluding in its paradoxical language to dimensions of reality that we instinctively relate to ourselves.

Al Zolynas, in his prose poem "The New Physics," describes the precarious balance on this "see-saw of paradox." In his metaphors the physicist is described as a man at serious play. He is a man of gravity, a juggler, a tumbler, a puzzle solver in search of stable bearings in this infinitely receding universe. Yet he is at home, happy, part of a family engaging in its magic, which is to him no less wondrous. He is, in short, like the particle in the box, in two places at the same time. The loss of a universal hitching post implied by relativity theory has not, for the man of this world, undermined his basic human and moral values.

From a poet's point of view, relativity gave back, in new form, much of what Copernicus took away—the human center and the value of individual perception. This is best expressed in Einstein's demonstration of time and space as variables that can be meaningfully specified only with reference to an observer. In a pattern poem by David Pettys, for example, one of the infinite centers of this open-ended universe is perceived as the place under the table of a writer where his typewriter sits. Both space and time as relative coordinates are the basis for two of Marvin Cohen's idiosyncratic fables that generate wit from cosmic paradox. In a third piece, he imagines Picasso making his way to an eye doctor through the subjective space of his own creation. For all of these contemporary writers, the subjective nature of space and time reinforces belief in the individual life and its "relatively" important contribution to being.

Siv Cedering uses this idea to contemplate the intellect of

Einstein himself. In her poem, "Letter from Albert Einstein (1879–1955)," the physicist is seen as reducing the immensity of space to communicable thought. Not every poet responds so positively to man's reduction of nature. For Richard Eberhart it implies the end of romantic vision. Angered by man "Shooting Particles beyond the World," he expresses his rage in a deliberate parody of Wordsworth's romanic language. During the age of industrial fortune seeking, Wordsworth had observed with some disdain:

> The world is too much with us; late and soon,
> Getting and spending, we lay waste our powers:

Eberhart, removing himself from association with particle-shooting man, lashes him with Wordsworth's altered line:

> The world is too much for *him*. The green
> Of earth is not enough, love's deities.

The heart of the argument is the same in both poems—that man has laid waste to the planet and killed its myths; the difference is that Wordsworth saw evidence of this only in cities, while Eberhart to his great disgust finds the whole earth in ruins. He calls man an "infant," "maniac," and, borrowing Sir Izaak Walton's seventeenth-century title, a "compleat angler." Like the fisherman, he is "casting" lures abroad, though his "slugs," the poem argues, merely "flaunt his out-cast state" and "spit on the sun."

Eberhart's vision of degraded man does, nevertheless, perpetuate a certain egotism on behalf of mankind. It implies that man still makes a difference even if the difference is limited to the power of injuring the universe. This is a presumptuous inference given the second law of thermodynamics, which proclaims the irreversible process of nature that seems to be moving in the direction of maximum entropy. The universe is running down. With or without man's wounding particles, the sun is slowly burning out; matter and energy are inexorably scattering into a space that grows cold and empty. Yet for most people the ultimate future projected by theoretical physics is less fearful than the immediate future described by Eberhart. His rage has humanistic motives—to slow down human genocide. There is no doubt that the formidable equivalence of mass and energy has given mankind the potential for immediate annihilation and that $E = mc^2$ hovers over us like a contraction of Dante's message, "Abandon hope all ye who enter here."

Atomic physics has come to epitomize human experience in this age. It both accounts for and expresses our loss of scale, our fragmentation, subjectivity, and sense of void. Yet poets are reluctant to abandon hope. By altering our world view, Josephine Miles concludes, physics has also inadvertently fostered a powerful vocabulary for probing the ultimate form and mystery that underlie art and poetry, no less than the world itself.

PLINY THE ELDER

(23–79)

From *Natural History*

To pass on to the other more remarkable stones, who can for a moment doubt that the magnet will be the first to suggest itself? For what, in fact, is there endowed with more marvellous properties than this? or in which of her departments has Nature displayed a greater degree of waywardness? She had given a voice to rocks, as already mentioned, and had enabled them to answer man, or rather, I should say, to throw back his own words in his teeth. What is there in existence more inert than a piece of rigid stone? And yet, behold! Nature has here endowed stone with both sense and hands. What is there more stubborn than hard iron? Nature has, in this instance, bestowed upon it both feet and intelligence. It allows itself, in fact, to be attracted by the magnet, and, itself a metal which subdues all other elements, it precipitates itself towards the source of an influence at once mysterious and unseen. The moment the metal comes near it, it springs towards the magnet, and, as it clasps it, is held fast in the magnet's embraces. Hence it is that this stone is sometimes known by the name of "sideritis" [iron earth]; another name given to it being "heraclion" [after Hercules]. It received its name "magnes," Nicander informs us, from the person who was the first to discover it, upon Ida. It is found, too, in various other countries, as in Spain for example. Magnes, it is said, made this discovery, when, upon taking his herds to pasture, he found that the nails of his shoes and the iron ferrel of his staff adhered to the ground.

Sotacus describes five different kinds of magnet; the Æthiopian magnet; that of Magnesia, a country which borders on Macedonia, and lies to the right of the road which leads from the town of Boebe to Iolcos; a third, from Hyettus in Boeotia; a fourth, from Alexandria in Troas; and a fifth, from Magnesia in Asia. The leading distinction in magnets is the sex, male and female, and the next great difference in them is the colour. Those of Magnesia, bordering on Macedonia, are of a reddish black; those of Boeotia are more red than black; and the kind that is found in Troas is black, of the female sex, and consequently destitute of attractive power. The most inferior, however, of all, are those of Magnesia in Asia: they are white, have no attractive influence on iron, and resemble pumice in appearance. It has been found by experience, that the more nearly the magnet approaches to an azure colour, the better it is in quality. The Æthiopian magnet is looked upon as the best of all, and is purchased at its weight in silver: Zmiris in Æthiopia is the place where it is found, such being the name of a region there, covered with sand.

In the same country, too, the magnet called "haematites" is found, a stone of a blood-red colour, and which, when bruised, yields a tint like that of blood, as also of saffron. The haematites has not the same property of attracting iron that the ordinary magnet has. The Æthiopian magnet is recognized by this peculiarity, that it has the property, also, of attracting other magnets to it. All these minerals are useful as ingredients in ophthalmic preparations, in certain proportions according to the nature of each: they are particularly good, too, for arresting defluxions of the eyes. Triturated in a calcined state, they have a healing effect upon burns.

PETRUS PEREGRINUS
[PIERRE DE MARICOURT]
(fl. 13th cent. A.D.)

From *Letter on the Magnet*

Dearest of Friends:

At your earnest request, I will now make known to you, in an unpolished narrative, the undoubted though hidden virtue of the lodestone, concerning which philosophers up to the present time give us no information, because it is characteristic of good things to be hidden in darkness until they are brought to light by application to public utility. Out of affection for you, I will write in a simple style about things entirely unknown to the ordinary individual. Nevertheless I will speak only of the manifest properties of the lodestone, because this tract will form part of a work on the construction of philosophical instruments. The disclosing of the hidden properties of this stone is like the art of the sculptor by which he brings figures and seals into existence. Although I may call the matters about which you inquire evident and of inestimable value, they are considered by common folk to be illusions and mere creations of the imagination. But the things that are hidden from the multitude will become clear to astrologers and students of nature, and will constitute their delight, as they will also be of great help to those that are old and more learned.

CHAPTER II
QUALIFICATIONS OF THE EXPERIMENTER

You must know, my dear friend, that who-ever wishes to experiment, should be acquainted with the nature of things, and should not be ignorant of the motion of the celestial bodies. He must also be skilful in manipulation in order that, by means of this stone, he may produce these marvelous effects. Through his own industry he can, to some extent, indeed, correct the errors that a mathematician would inevitably make if he were lacking in dexterity. Besides, in such occult experimentation, great skill is required, for very frequently without it the desired result cannot be obtained, because there are many things in the domain of reason which demand this manual dexterity.

*   *   *

CHAPTER IV
HOW TO DISTINGUISH THE POLES OF A LODESTONE

I wish to inform you that this stone bears in itself the likeness of the heavens, as I will now clearly demonstrate. There are in the heavens two points more important than all others, because on them, as on pivots, the celestial sphere revolves: these points are called, one the arctic or north pole, the other the antarctic or south pole. Similarly you must fully realize that in this stone there are two points styled respectively the north pole and the south pole. If you are very careful, you can discover these two points in a general way. One method for doing so is the following: With an instrument with which crystals and other stones are rounded let a lodestone be made into a globe and then polished. A needle or an elongated piece of iron is then placed on top of the lodestone and a line is drawn in the direction of the needle or iron, thus dividing the stone into two equal parts. The needle is next placed on another part of the stone and a second median line drawn. If desired, this operation may be performed on many different parts, and undoubtedly all these lines will meet in two points just as all meridian or azimuth circles meet in the two opposite poles of the globe. One of these is the north pole, the other the south pole. . . .

A second method for determining these important points is this: Note the place on the above-mentioned spherical lodestone where the point of the needle clings most frequently and most strongly; for this will be one of the poles as discovered by the previous method. In order to determine this point exactly, break off a small piece of the needle or iron so as to obtain a fragment about the length of two fingernails; then put it on the spot which was found to be the pole by the former operation. If the fragment stands perpendicular to the stone, then that is, unquestionably, the pole sought; if not, then move the iron fragment about until it becomes so; mark this point carefully; on the opposite end another point may be found in a similar manner. If all this has been done rightly, and if the stone is homogeneous throughout and a choice specimen, these two points will be diametrically opposite, like the poles of a sphere.

*   *   *

CHAPTER VI
HOW ONE LODESTONE ATTRACTS ANOTHER

When you have discovered the north and the south pole in your lodestone, mark them both carefully, so that by means of these indentations they may be distinguished whenever necessary. Should you wish to see how one lodestone attracts another, then, with two lodestones selected and prepared as mentioned. . . , proceed as follows: Place one in its dish that it may float about as a sailor in a skiff, and let its poles which have already been determined be equidistant from the horizon, i.e., from the edge of the vessel. Taking the other stone in your hand, approach its north pole to the south pole of the lodestone floating in the vessel; the latter will follow the stone in your hand as if longing to cling to it. If, conversely, you bring the south end of the lodestone in your hand toward the north end of the floating lodestone, the same phenomenon will occur; namely, the floating lodestone will follow the one in your hand. Know then that this is the law; the north pole of one lodestone attracts the south pole of another, while the south pole attracts the north. Should you proceed otherwise and bring the north pole of one near the north pole of another, the one you hold in your hand will seem to put the floating one to flight. If the south pole of one is brought near the south pole of another, the same will happen. This is because the north pole of one seeks the south pole of the other, and therefore repels the north pole. A proof of this is that finally the north pole becomes united with the south pole. Likewise if the south pole is stretched out towards the south pole of the floating lodestone, you will observe the latter to be repelled, which does not occur, as said before, when the north pole is extended towards the south. Hence the silliness of certain persons is manifest, who claim that just as scammony[1] attracts jaundice on account of a similarity between them, so one lodestone attracts another even more strongly than it does iron, a fact which they suppose to be false although really true as shown by experiment.

*   *   *

CHAPTER VIII
HOW A LODESTONE ATTRACTS IRON

If you wish the stone, according to its natural desire, to attract iron, proceed as follows: Mark the north end of the iron and towards this end approach the south pole of the stone, when it will be found to follow the latter. Or, on the contrary, to the south part of the iron present the north pole of the stone and the latter will attract it without any difficulty. Should you, however, do the opposite, namely, if you bring the north end of the stone towards the north pole of the iron, you will notice

[1] An herb.

the iron turn round until its south pole unites with the north end of the lodestone. The same thing will occur when the south end of the lodestone is brought near the south pole of the iron. Should force be exerted at either pole, so that when the south pole of the iron is made [to] touch the south end of the stone, then the virtue in the iron will be easily altered in such a manner that what was before the south end will now become the north and conversely. The cause is that the last impression acts, confounds, or counteracts and alters the force of the original movement.

* * *

CHAPTER X
AN INQUIRY INTO THE CAUSE OF THE NATURAL VIRTUE OF THE LODESTONE

Certain persons who were but poor investigators of nature held the opinion that the force with which a lodestone draws iron, is found in the mineral veins themselves from which the stone is obtained; whence they claim that the iron turns towards the poles of the earth, only because of the numerous iron mines found there. But such persons are ignorant of the fact that in many different parts of the globe the lodestone is found; from which it would follow that the iron neeedle should turn in different directions according to the locality; but this is contrary to experience. Secondly, these individuals do not seem to know that the places under the poles are uninhabitable because there one-half the year is day and the other half night. Hence it is most silly to imagine that the lodestone should come to us from such places. Since the lodestone points to the south as well as to the north, it is evident from the foregoing chapters that we must conclude that not only from the north pole but also from the south pole rather than from the veins of the mines virtue flows into the poles of the lodestone. This follows from the consideration that wherever a man may be, he finds the stone pointing to the heavens in accordance with the position of the meridian; but all meridians meet in the poles of the world; hence it is manifest that from the poles of the world, the poles of the lodestone receive their virtue. Another necessary consequence of this is that the needle does not point to the pole star, since the meridians do not intersect in that star but in the poles of the world. In every region, the pole star is always found outside the meridian except twice in each complete revolution of the heavens. From all these considerations, it is clear that the poles of the lodestone derive their virtue from the poles of the heavens.

THOMAS BROWNE
(1605–82)

From *Pseudodoxia Epidemica*

And first we conceive the earth to be a Magneticall body. A Magneticall body, we term not only that which hath a power attractive; but that which sweated in a convenient medium naturally disposeth it self to one invariable and fixed situation. And such a Magneticall vertue we conceive to be in the Globe of the earth; whereby as unto its naturall points and proper terms it disposeth it self unto the poles; being so framed, constituted, and ordered unto these points, that those parts which are now at the poles, would not naturally abide under the Aequator; nor Green-land remain in the place of Magellanica. And if the whole earth were violently removed, yet would it not forgoe its primitive points; nor pitch in the East or West, but return unto its polary position again. For though by compactnesse or gravity it may acquire the lowest place, and become the center of the universe, yet that it makes good that point, not varying at all by the accession of bodies upon, or secession thereof, from its surface, perturbing the equilibration of either Hemisphere (whereby the altitude of the starres might vary) or that it strictly maintains the north and southern points; that neither upon the motions of the heavens, ayre, and windes without, large eruptions and division of parts within, its polary parts should never incline or veere unto the Aequator (whereby the latitude of places should also vary) it cannot so well be salved from gravity as a Magneticall verticity. This is probably that foundation the wisdome of the Creator hath laid unto the earth; in this sense we may more nearly apprehend, and sensibly make out the expressions of holy Scripture, as that of Psal. 93.1. *Firmavit orbem terrae qui non commovebitur*, he hath made the round world so sure that it cannot be moved: as when it is said by Job, *Extendit Aquilonem super vacuo, &c.* He stretcheth forth the North upon the empty place, and hangeth the earth upon nothing. And this is the most probable answer unto that great question, Job 38. Whereupon are the foundations of the earth fastened, or who laid the corner stone thereof? Had they been acquainted with this principle, Anaxagoras, Socrates, and Democritus had better made out the ground of this stability: Xenophanes had not been fain to say the earth had no bottome; and Thales Milesius to make it swim in water. (Now whether the earth stand still, or moveth circularly, we may concede this Magneticall stability: For although it move, in that conversion the poles and center may still remaine the same, as is conceived in the Magneticall bodies of heaven, especially Jupiter and the Sunne; which according to Galileus, Kepler, and Fabricius, are observed to have Dineticall motions and certaine revolutions about the proper centers; and though the one in about the space of ten dayes, the other in less than one accomplish this revolution, yet do they observe a constant habitude unto their poles and firme themselves thereon in their gyration.)

Nor is the vigour of this great body included only in its self, or circumferenced by its surface, but diffused at indeterminate distances through the ayre, water, and all bodies circumjacent; exciting and impregnating Magneticall bodies within its surface or without it, and performing in a secret and invisible way what we evidently behold effected by the Loadstone. For these effluxions penetrate all bodies, and like the species of visible objects are ever ready in the medium, and lay hold on all bodies proportionate or capable of their action;

those bodies likewise being of a congenerous nature doe readily receive the impressions of their motor; and if not fettered by their gravity, conform themselves to situations, wherein they best unite unto their Animator. And this will sufficiently appear from the observations that are to follow, which can no better way be made out than this we speak of, the Magneticall vigour of the earth. Now whether these effluviums do flye by striated Atomes and winding particles as *Renatus des Cartes*[2] conceiveth, or glide by streams attracted from either pole and Hemisphere of the earth unto the Aequator, as Sir Kenelme Digby[3] excellently declareth, it takes not away this vertue of the earth; but more distinctly sets down the gests and progresse thereof; and are conceits of eminent use to salve Magneticall phenomenas. And as in Astronomy those hypotheses though never so strange are best esteemed which best doe salve apparencies; so surely in Philosophy those principles (though seeming monstrous) may with advantage be embraced, which best confirm experiment, and afford the readiest reason of observation. And truly the doctrine of effluxions, their penetrating natures, their invisible paths, and unsuspected effects, are very considerable; for besides this Magneticall one of the earth, severall effusions there may be from divers other bodies, which invisibly act their parts at any time, and perhaps through any medium; a part of Philosophy but yet in discovery, and will I fear prove the last leaf to be turned over in the book of Nature.[4]

[2] René Descartes (1596–1650), French mathematician.
[3] Digby (1603–65), a poet, philosopher, bibliophile, and a master chef.
[4] He goes on to discuss the experiments of William Gilbert, who published *De Magnete* in 1600. These experiments involved heating, chilling, and floating lodestones.

FRANCIS QUARLES
(1592–1644)

From *Emblems*

BOOK V, NO. IV

1   Like to the arctic needle, that doth guide
      The wand'ring shade by his magnetic pow'r,
    And leaves his silken gnomon to decide
      The question of the controverted hour,
    First frantics up and down from side to side,
      And restless beats his crystall'd iv'ry case,
      With vain impatience jets from place to place,
    And seeks the bosom of his frozen bride;
      At length he slacks his motion, and doth rest
    His trembling point at his bright pole's beloved breast.

2   E'en so my soul, being hurried here and there,
      By ev'ry object that presents delight,
    Fain would be settled, but she knows not where;
      She likes at morning what she loathes at night:
    She bows to honour; then she lends an ear
      To that sweet swan-like voice of dying pleasure;
      Then tumbles in the scatter'd heaps of treasure;
    Now flatter'd with false hopes; now foil'd with fear:
      Thus finding all the world's delight to be
    But empty toys, good God! she points alone to thee.

3   But hath the virtued steel a power to move?
      Or can the untouch'd needle point aright?
    Or can my wand'ring thoughts forbear to rove,

Unguided by the virtue of thy Sp'rit?
O hath my leaden soul the art t' improve
   Her wasted talent, and, unraised, aspire
   In this sad moulting time of her desire?
Not first beloved, have I the power to love?
   I cannot stir, but as thou please to move me,
Nor can my heart return thee love, until thou love me.

4   The still commandress of the silent night
   Borrows her beams from her bright brother's eye;
His fair aspect fills her sharp horns with light,
   If he withdraw, her flames are quench'd and die:
E'en so the beams of thy enlight'ning Sp'rit,
   Infused and shot into my dark desire,
   Inflame my thoughts, and fill my soul with fire,
That I am ravish'd with a new delight;
   But if thou shroud thy face, my glory fades,
And I remain a nothing, all composed of shades.

5   Eternal God! O thou that only art
   The sacred fountain of eternal light,
And blessed loadstone of my better part,
   O thou, my heart's desire, my soul's delight!
Reflect upon my soul, and touch my heart,
   And then my heart shall prize no good above thee;
   And then my soul shall know thee; knowing, love
     thee;
And then my trembling thoughts shall never start
   From thy commands, or swerve the least degree,
Or once presume to move, but as they move in thee.

THOMAS STANLEY
(1625–78)

The Magnet

Ask the Empress of the night
  How the hand which guides her sphere,
Constant in unconstant light,
  Taught the waves her yoke to bear,
And did thus by loving force
Curb or tame the rude seas course.

Ask the female Palme how shee
  First did woo her husbands love;
And the Magnet, ask how he
  Doth th'obsequious iron move;
Waters, plants and stones know this,
That they love, not what love is.

Be not then less kind than these,
  Or from love exempt alone,
Let us twine like amorous trees,
  And like rivers melt in one;
Or if thou more cruell prove
Learne of steel and stones to love.

HENRY VAUGHAN
(1622–95)

The Starre

What ever 'tis whose beauty here below
Attracts thee thus & makes thee stream & flow,
    And wind and curle, and wink and smile,
      Shifting thy gate and guile:

Though thy close commerce nought at all imbarrs
My present search, for eagles eye not starrs,
    And still the lesser by the best
      And highest good is blest:

Yet, seeing all things that subsist and be,
Have their Commissions from Divinitie,
    And teach us duty, I will see
      What man may learn from thee.

First, I am sure, the Subject so respected
Is well disposed, for bodies once infected,
    Deprav'd or dead, can have with thee
      No hold, nor sympathie.

Next, there's in it a restless, pure desire
And longing for thy bright and vitall fire,
    Desire that never will be quench'd
      Nor can be writh'd, nor wrench'd.

These are the Magnets which so strongly move
And work all night upon thy light and love,
    As beauteous shapes, we know not why,
      Command and guide the eye.

For where desire, celestiall, pure desire
Hath taken root, and grows, and doth not tire,
    There God a Commerce states, and sheds
      His secret on their Heads.

This is the Heart he craves; and who so will
But give it him, and grudge not; he shall feel
    That God is true, as herbs unseen
      Puts on their youth and green.

JONATHAN SWIFT
(1667–1745)

From *Gulliver's Travels*

A Phenomenon Solved by Modern Philosophy and Astronomy. The Laputians' Great Improvements in the Latter. The King's Method of Suppressing Insurrections.

I desired leave of this prince to see the curiosities of the island, which he was graciously pleased to grant, and ordered my tutor to attend me. I chiefly wanted to know to what cause in art or in nature it owed its several motions, whereof I will now give a philosophical account to the reader.

The Flying or Floating Island is exactly circular, its diameter 7,837 yards, or about four miles and an half, and consequently contains ten thousand acres. It is three hundred yards thick. The bottom or under surface, which appears to those who view it from below, is one even regular plate of adamant, shooting up to the height of about two hundred yards. Above it lie the several minerals in their usual order, and over all is a coat of rich mould ten or twelve foot deep. The declivity of the upper surface, from the circumference to the center, is the natural cause why all the dews and rains which fall upon the island are conveyed in small rivulets towards the middle, where they are emptied into four large basons, each of about half a mile in circuit, and two hundred yards distant from the center. From these basons the water is continually exhaled by the sun in the day time, which effectually prevents their overflowing. Besides, as it is in the power of the monarch to raise the island above the region of clouds and vapours, he can prevent the falling of dews and rains when ever he pleases. For the highest clouds cannot rise above two miles, as naturalists agree, at least they were never known to do so in that country.

At the center of the island there is a chasm about fifty yards in diameter, from whence the astronomers descend into a large dome, which is therefore called *Flandona Gagnole,* or the *Astronomer's Cave,* situated at the depth of an hundred yards beneath the upper surface of the adamant. In this cave are twenty lamps continually burning, which from the reflection of the adamant cast a strong light into every part. The place is stored with great variety of sextants, quadrants, telescopes, astrolabes, and other astronomical instruments. But the greatest curiosity, upon which the fate of the island depends, is a loadstone of a prodigious size, in shape resembling a weaver's shuttle. It is in length six yards, and in the thickest part at least three yards over. This magnet is sustained by a very strong axle of adamant passing through its middle, upon which it plays, and is poised so exactly that the weakest hand can turn it. It is hooped round with an hollow cylinder of adamant, four foot deep, as many thick, and twelve yards in diameter, placed horizontally, and supported by eight adamantine feet, each six yards high. In the middle of the concave side there is a groove twelve inches deep, in which the extremities of the axle are lodged, and turned round as there is occasion.

The stone cannot be moved from its place by any force, because the hoop and its feet are one continued piece with that body of adamant which constitutes the bottom of the island.

By means of this loadstone, the island is made to rise and fall, and move from one place to another. For, with respect to that part of the earth over which the monarch presides, the stone is endued at one of its sides with an attractive power, and at the other with a repulsive. Upon placing the magnet erect with its attracting end towards the earth, the island descends; but when the repelling extremity points downwards, the island mounts directly upwards. When the position of the stone is oblique, the motion of the island is so too. For in this magnet the forces always act in lines parallel to its direction.

By this oblique motion the island is conveyed to different parts of the monarch's dominions. To explain the manner of its progress, let *A B* represent a line drawn cross the dominions of Balnibarbi, let the line *c d* represent the loadstone, of which let *d* be the repelling end, and *c* the attracting end, the island being over *C*; let the stone be placed in the position *c d* with its repelling end downwards; then the island will be driven upwards obliquely towards *D*. When it is arrived at *D*, let the stone be turned upon its axle till its attracting end points towards *E*, and then the island will be carried obliquely towards *E*; where if the stone be again turned upon its axle till it stands in the position *E F*, with its repelling point downwards, the island will rise obliquely towards *F*, where by directing the attracting end towards *G*, the island may be carried to *G*, and from *G* to *H*, by turning the stone, so as to make its repelling extremity point directly downwards. And thus by changing the situation of the stone as often as there is occasion, the island is made to rise and fall by turns in an oblique direction, and by those alternate risings and fallings (the obliquity being not considerable) is conveyed from one part of the dominions to the other.

But it must be observed, that this island cannot move beyond the extent of the dominions below, nor can it rise above the height of four miles. For which the astronomers (who have written large systems concerning the stone) assign the following reason: that the magnetic virtue does not extend beyond the distance of four miles, and that the mineral which acts upon the stone in the bowels of the earth, and in the sea about six leagues distant from the shore, is not diffused through the whole globe, but terminated with the limits of the King's dominions; and it was easy, from the great advantage of such a superior situation, for a prince to bring under his obedience whatever country lay within the attraction of that magnet.

When the stone is put parallel to the plane of the horizon, the island standeth still; for in that case, the extremities of it, being at equal distance from the earth, act with equal force, the one in drawing downwards, the other in pushing upwards, and consequently no motion can ensue.

This loadstone is under the care of certain astronomers, who from time to time give it such positions as the monarch directs. They spend the greatest part of their lives in observing the celestial bodies, which they do by the assistance of glasses far excelling ours in goodness. For although their largest telescopes do not exceed three feet, they magnify much more than those of a hundred with us, and at the same time show the stars with greater clearness. For this advantage hath enabled them to extend their discoveries much farther than our astronomers in Europe. They have made a catalogue of ten thousand fixed stars, whereas the largest of ours do not contain above one third part of that number. They have likewise discovered two lesser stars, or "satellites," which revolve about Mars, whereof the innermost is distant from the center of the primary planet exactly three of his diameters, and the outermost five; the former revolves in the space of ten hours, and the latter in twenty-one and an half; so that the squares of their periodical times are very near in the same proportion with the cubes of their distance from the center of Mars, which evidently shows them to be governed by the same law of gravitation, that influences the other heavenly bodies.

They have observed ninety-three different comets, and settled their periods with great exactness. If this be true (and they affirm it with great confidence) it is much to be wished that their observations were made public, whereby the theory of comets, which at present is very lame and defective, might be brought to the same perfection with other parts of astronomy.

The King would be the most absolute prince in the universe, if he could but prevail on a ministry to join with him; but these having their estates below on the continent, and considering that the office of a favourite hath a very uncertain tenure, would never consent to the enslaving their country.

If any town should engage in rebellion or mutiny, fall into violent factions, or refuse to pay the usual tribute, the King hath two methods of reducing them to obedience. The first and the mildest course is by keeping the island hovering over such a town, and the lands about it, whereby he can deprive them of the benefit of the sun and the rain, and consequently afflict the inhabitants with dearth and diseases. And if the crime deserve it they are at the same time pelted from above with great stones, against which they have no defence but by creeping into cellars or caves, while the roofs of their houses are beaten to pieces. But if they still continue obstinate, or offer to raise insurrections, he proceeds to the last remedy, by letting the island drop directly upon their heads, which makes a universal destruction both of houses and men.

FRANCIS QUARLES

(1592–1644)

From *Emblems*

BOOK III, NO. XIV

FLESH. What means my sister's eye so oft to pass
Through the long entry of that optic-glass?
Tell me; what secret virtue doth invite
Thy wrinkled eye to such unknown delight?

SPIRIT.　It helps the sight, makes things remote appear
　　　　In perfect view; it draws the objects near
FLESH.　What sense-delighting objects dost thou spy?
　　　　What doth that glass present before thine eye?
SPIRIT.　I see thy foe, my reconciled friend,
　　　　Grim Death, e'en standing at the glass's end:
　　　　His left hand holds a branch of palm; his right
　　　　Holds forth a two-edg'd sword. FLESH. A proper
　　　　　sight.
　　　　And is this all? doth thy prospective please
　　　　Th' abused fancy with no shapes but these?
SPIRIT.　Yes, I behold the darken'd sun bereav'n
　　　　Of all his light, the battlements of Heav'n
　　　　Swelt'ring in flames; the angel-guarded Son
　　　　Of glory on his high tribunal-throne;
　　　　I see a brimstone sea of boiling fire,
　　　　And fiends, with knotted whips of flaming wire,
　　　　Tort'ring poor souls, that gnash their teeth in
　　　　　vain,
　　　　And gnaw their flame-tormented tongues for
　　　　　pain.
　　　　Look, sister, how the queasy-stomach'd graves
　　　　Vomit their dead, and how the purple waves
　　　　Scald their consumeless bodies, strongly cursing
　　　　All wombs for bearing, and all paps for nursing.
FLESH.　Can thy distemper'd fancy take delight
　　　　In view of tortures? these are shows t' affright:
　　　　Look in this glass triangular; look here,
　　　　Here's that will ravish eyes. SPIRIT. What seest
　　　　　thou there?
FLESH.　The world in colours; colours that distain
　　　　The cheeks of Proteus, or the silken train
　　　　Of Flora's nymphs; such various sorts of hue,
　　　　As sun-confronting Iris never knew:
　　　　Here, if thou please to beautify a town,
　　　　Thou may'st; or with a hand, turn't upside down;
　　　　Here may'st thou scant or widen by the measure
　　　　Of thine own will; make short or long at pleasure:
　　　　Here may'st thou tire thy fancy, and advise
　　　　With shows more apt to please more curious eyes.

SPIRIT.　Ah, fool! that doat'st on vain, on present toys,
And disrespect'st those true, those future joys:
How strongly are thy thoughts befool'd, alas!
To doat on goods that perish with thy glass!
Nay, vanish with the turning of a hand:
Were they but painted colours, it might stand
With painted reason that they might devote thee;
But things that have no being to besot thee!
Foresight of future torments is the way
To balk those ills which present joys betray.
As thou hast fool'd thyself, so now come hither,
Break that fond glass, and let's be wise together.

JAMES THOMSON
(1700–48)

To the Memory of Sir Isaac Newton

Shall the great soul of Newton quit the earth
To mingle with his stars, and every Muse,
Astonished into silence, shun the weight
Of honours due to his illustrious name?
But what can man? Even now the sons of light,
In strains high warbled to seraphic lyre,
Hail his arrival on the coast of bliss.
Yet am not I deterred, though high the theme.
And sung to harps of angels for with you,
Ethereal flames! ambitious, I aspire
In Nature's general symphony to join.
　　And what new wonders can ye show your guest!
Who, while on this dim spot where mortals toil

Clouded in dust, from motion's simple laws
Could trace the secret hand of Providence,
Wide-working through this universal frame.
   Have ye not listened while he bound the suns
And planets to their spheres! the unequal task
Of humankind till then. Oft had they rolled
O'er erring man the year, and oft disgraced
The pride of schools, before their course was known
Full in its causes and effects to him,
All-piercing sage! who sat not down and dreamed
Romantic schemes, defended by the din
Of specious words, and tyranny of names;
But, bidding his amazing mind attend,
And with heroic patience years on years
Deep-searching, saw at last the system dawn,
And shine, of all his race, on him alone.
   What were his raptures then! how pure! how strong!
And what the triumphs of old Greece and Rome,
By his diminished, but the pride of boys
In some small fray victorious! when instead
Of shattered parcels of this earth usurped
By violence unmanly, and sore deeds
Of cruelty and blood, Nature herself
Stood all subdued by him, and open laid
Her every latent glory to his view.
   All intellectual eye, our solar round
First gazing through, he, by the blended power
Of gravitation and projection, saw
The whole in silent harmony revolve.
From unassisted vision hid, the moons
To cheer remoter planets numerous formed,
By him in all their mingled tracts were seen.
He also fixed our wandering Queen of Night,
Whether she wanes into a scanty orb,
Or, waxing broad, with her pale shadowy light,
In a soft deluge overflows the sky.
Her every motion clear-discerning, he
Adjusted to the mutual main and taught
Why now the mighty mass of waters swells

Resistless, heaving on the broken rocks,
And the full river turning—till again
The tide revertive, unattracted, leaves
A yellow waste of idle sands behind.
  Then, breaking hence he took his ardent flight
Through the blue infinite; and every star,
Which the clear concave of a winter's night
Pours on the eye, or astronomic tube,
Far stretching, snatches from the dark abyss,
Or such as further in successive skies
To fancy shine alone, at his approach
Blazed into suns, the living centre each
Of an harmonious system—all combined,
And ruled unerring by that single power
Which draws the stone projected to the ground.
  O unprofuse magnificence divine!
O wisdom truly perfect! thus to call
From a few causes such a scheme of things,
Effects so various, beautiful, and great,
An universe complete! And O beloved
Of Heaven! whose well purged penetrating eye
The mystic veil transpiercing, inly scanned
The rising, moving, wide-established frame.
  He, first of men, with awful wing pursued
The comet through the long elliptic curve,
As round innumerous worlds he wound his way,
Till, to the forehead of our evening sky
Returned, the blazing wonder glares anew,
And o'er the trembling nations shakes dismay.
  The heavens are all his own, from the wide rule
Of whirling vortices and circling spheres
To their first great simplicity restored.
The schools astonished stood; but found it vain
To combat still with demonstration strong,
And, unawakened, dream beneath the blaze
Of truth. At once their pleasing visions fled,
With the gay shadows of the morning mixed,
When Newton rose, our philosophic sun!
  The aerial flow of sound was known to him,

From whence it first in wavy circles breaks,
Till the touched organ takes the message in.
Nor could the darting beam of speed immense
Escape his swift pursuit and measuring eye.
Even Light itself, which every thing displays,
Shone undiscovered, till his brighter mind
Untwisted all the shining robe of day;
And, from the whitening undistinguished blaze,
Collecting every ray into his kind,
To the charmed eye educed the gorgeous train
Of parent colours. First the flaming red
Sprung vivid forth; the tawny orange next;
And next delicious yellow; by whose side
Fell the kind beams of all-refreshing green.
Then the pure blue, that swells autumnal skies,
Ethereal played; and then, of sadder hue,
Emerged the deepened indigo, as when
The heavy-skirted evening droops with frost;
While the last gleamings of refracted light
Died in the fainting violet away.
These, when the clouds distil the rosy shower,
Shine out distinct adown the watery bow;
While o'er our heads the dewy vision bends
Delightful, melting on the fields beneath.
Myriads of mingling dyes from these result,
And myriads still remain—infinite source
Of beauty, ever flushing, ever new.
    Did ever poet image aught so fair,
Dreaming in whispering groves by the hoarse brook?
Or prophet, to whose rapture heaven descends?
Even now the setting sun and shifting clouds,
Seen, Greenwich, from thy lovely heights, declare
How just, how beauteous the refractive law.
    The noiseless tide of time, all bearing down
To vast eternity's unbounded sea,
Where the green islands of the happy shine,
He stemmed alone; and, to the source (involved
Deep in primeval gloom) ascending, raised
His lights at equal distances, to guide

Historian wildered on his darksome way.
　　But who can number up his labours? who
His high discoveries sing? When but a few
Of the deep-studying race can stretch their minds
To what he knew—in fancy's lighter thought
How shall the muse then grasp the mighty theme?
　　What wonder thence that his devotion swelled
Responsive to his knowledge? For could he
Whose piercing mental eye diffusive saw
The finished university of things
In all its order, magnitude, and parts
Forbear incessant to adore that Power
Who fills, sustains, and actuates the whole?
　　Say, ye who best can tell, ye happy few,
Who saw him in the softest lights of life,
All unwithheld, indulging to his friends
The vast unborrrowed treasures of his mind,
Oh, speak the wondrous man! how mild, how calm,
How greatly humble, how divinely good,
How firmly stablished on eternal truth;
Fervent in doing well, with every nerve
Still pressing on, forgetful of the past,
And panting for perfection; far above
Those little cares and visionary joys
That so perplex the fond impassioned heart
Of ever cheated, ever trusting man
This, Conduitt, from thy rural hours we hope,
As through the pleasing shade where nature pours
Her every sweet in studious ease you walk,
The social passions smiling at thy heart
That glows with all the recollected sage.
　　And you, ye hopeless gloomy-minded tribe,
You who, unconscious of those nobler flights
That reach impatient at immortal life
Against the prime endearing privilege
Of being dare contend,—say, can a soul
Of such extensive, deep, tremendous powers,
Enlarging still, be but a finer breath
Of spirits dancing through their tubes awhile,

And then for ever lost in vacant air?
   But hark! methinks I hear a warning voice,
Solemn as when some awful change is come,
Sound through the world—"Tis done!—the measure's full;
And I resign my charge."—Ye mouldering stones
That build the towering pyramid, the proud
Triumphal arch, the monument effaced
By ruthless ruin, and whate'er supports
The worshipped name of hoar antiquity—
Down to the dust! What grandeur can ye boast
While Newton lifts his column to the skies,
Beyond the waste of time. Let no weak drop
Be shed for him. The virgin in her bloom
Cut off, the joyous youth, and darling child—
These are the tombs that claim the tender tear
And elegiac song. But Newton calls
For other notes of gratulation high,
That now he wanders through those endless worlds
He here so well descried, and wondering talks,
And hymns their Author with his glad compeers.
O Britain's boast! whether with angels thou
Sittest in dread discourse, or fellow-blessed,
Who joy to see the honour of their kind;
Or whether, mounted on cherubic wing,
Thy swift career is with the whirling orbs,
Comparing things with things, in rapture lost,
And grateful adoration for that light
So plenteous rayed into thy mind below
From Light Himself; oh, look with pity down
On humankind, a frail erroneous race!
Exalt the spirit of a downward world!
O'er thy dejected country chief preside,
And he her Genius called! her studies raise,
Correct her manners, and inspire her youth;
For, though depraved and sunk she brought thee forth,
And glories in thy name! she points thee out
To all her sons, and bids them eye thy star:
While, in expectance of the second life
When time shall be nor more, thy sacred dust
Sleeps with her kings, and dignifies the scene.

WILLIAM BLAKE

(1757–1827)

Miscellaneous Epigrams and Fragments

You don't believe—I won't attempt to make ye:
You are asleep—I won't attempt to wake ye.
Sleep on, Sleep on! while in your pleasant dreams
Of Reason you may drink of Life's clear streams.
Reason and Newton, they are quite two things;
For so the Swallow & the Sparrow sings.
Reason says "Miracle": Newton says "Doubt."
Aye! that's the way to make all Nature out.
"Doubt, Doubt, & don't believe without experiment":
That is the very thing that Jesus meant,
When he said, "Only Believe! Believe & try!
Try, Try, and never mind the Reason why."

Untitled

Mock on, Mock on Voltaire, Rousseau:
Mock on, Mock on: 'tis all in vain!
You throw the sand against the wind,
And the wind blows it back again.

And every sand becomes a Gem
Reflected in the beams divine;
Blown back they blind the mocking Eye,
But still in Israel's paths they shine.

The Atoms of Democritus
And Newton's particles of light
Are sands upon the Red sea shore,
Where Israel's tents do shine so bright.

WILLIAM COWPER

(1731–1800)

From "The Task"

Where finds the philosophy her eagle eye,
With which she gazes at yon burning disk
Undazzled, and detects and counts his spots?
In London: where her implements exact,
With which she calculates, computes, and scans,
All distance, motion, magnitude, and now
Measures an atom, and now girds a world?
In London.

*   *   *

God never meant that man should scale the heav'ns
By strides of human wisdom. In his works
Though wondrous, he commands us in his word
To seek *him* rather, where his mercy shines.
The mind indeed, enlighten'd from above,
Views him in all; ascribes to the grand cause
The grand effect; acknowledges with joy
His manner, and with rapture tastes his style.
But never yet did philosophic tube,
That brings the planets home into the eye
Of observation, and discovers, else
Not visible, his family of worlds,
Discover him that rules them; such a veil

Hangs over mortal eyes, blind from the birth,
And dark in things divine. Full often, too,
Our wayward intellect, the more we learn
Of nature, overlooks her author more;
From instrumental causes proud to draw
Conclusions retrograde, and mad mistake.
But if his word once teach us, shoot a ray
Through all the heart's dark chambers, and reveal
Truths undiscern'd but by that holy light,
Then all is plain. Philosophy, baptiz'd
In the pure fountain of eternal love,
Has eyes indeed; and viewing all she sees
As meant to indicate a God to man,
Gives *him* his praise, and forfeits not her own.
Learning has borne such fruit in other days
On all her branches: piety has found
Friends in the friends of science, and true pray'r
Has flow'd from lips wet with Castalian dews.
Such was thy wisdom, Newton, childlike sage!
Sagacious reader of the works of God,
And in his word sagacious.
                    [I, 712–19; III, 221–53]

CHARLES LAMB

(1775–1834)

From The Latin of Vincent Bourne[5]

NEWTON'S PRINCIPIA

Great Newton's self, to whom the world's in debt.
Owed to School Mistress sage his Alphabet;
But quickly wiser than his Teacher grown,
Discover'd properties to her unknown;
Of A *plus* B, or *minus*, learn'd the use,

[5]Vincent Bourne was William Cowper's teacher.

Known Quantities from unknown to educe;
And made—no doubt to that old dame's surprise—
The Christ-Cross-Row his Ladder to the skies.
Yet, whatsoe'er Geometricians say,
Her lessons were his true PRINCIPIA!

GEORGE GORDON, LORD BYRON

(1788–1824)

From *Don Juan*, Canto X

I

When Newton saw an apple fall, he found
  In that slight startle from his contemplation—
'Tis *said* (for I'll not answer above ground
  For any sage's creed or calculation)—
A mode of proving that the Earth turned round
  In a most natural whirl, called "gravitation";
And this is the sole mortal who could grapple,
Since Adam—with a fall—or with an apple.

II

Man fell with apples, and with apples rose,
  If this be true; for we must deem the mode
In which Sir Isaac Newton could disclose
  Through the then unpaved stars the turnpike road,
A thing to counterbalance human woes:
  For, ever since, immortal man hath glowed
With all kinds of mechanics, and full soon
Steam-engines will conduct him to the moon.

III

And wherefore this exordium?—Why, just now,
   In taking up this paltry sheet of paper,
My bosom underwent a glorious glow,
   And my internal spirit cut a caper:
And though so much inferior, as I know,
   To those who, by the dint of glass and vapour,
Discover stars, and sail in the wind's eye,
I wish to do as much by Poesy.

AL ZOLYNAS
(1945—)

The New Physics

—for Fritjof Capra

And so, the closer he looks at things, the farther away they
seem. At dinner, after a hard day at the universe, he finds
himself slipping through his food. His own hands wave at
him from beyond a mountain of peas. Stars and planets dance
with molecules on his fingertips. After a hard day with the
universe, he tumbles through himself, flies through the dream
galaxies of his own heart. In the very presence of his family he
feels he is descending through an infinite series of Chinese
boxes.

This morning, when he entered the little broom-closet of the
electron looking for quarks and neutrinos, it opened into an
immense hall, the hall into a plain—the Steppes of Mother
Russia! He could see men hauling barges up the river, chant-
ing faintly for their daily bread.

It's not that he longs for the old Newtonian Days, although something of plain matter and simple gravity might be reassuring, something of the good old equal-but-opposite forces. And it's not that he hasn't learned to balance comfortably on the see-saw of paradox. It's what he sees in the eyes of his children—the infinite black holes, the ransomed light at the center.

ERICA JONG

(1942—)

Half-Life

The rock I danced on
looked for all the world like the sea.
The sea was stone.

Your eyes were green
as wings of horseflies . . .
almost as unclean.

They buzzed around my head
like my own dreams
They thickened the air with kisses.

When I was nine.
I used to kiss my pillow
on the mouth
after I'd licked it wet.
How else find out
what "soul-kiss"
really meant?

"He puts his tongue
into your mouth"

I was amazed.

& yet our tongues are dancing
in the ocean.

Why does every fucking poem
mention the ocean?

The swell of the great sea mother?
The water babies in their amniotic fluid?
The sea salt taste of blood?

Love, blood—the flood of poems
as life creaks to a close.

The sky narrows to a point
as we make love.

This is a little death,
a pact,
a double suicide of sorts.

& I invent
tidal waves, atomic shocks,
the mushroom cloud of you
above the smoking chasm
that you leave in me.

Radioactive,
dangerous as stone,
you leave me bone dry, lonely in my cave.

I have compared you to atomic war.

& your half-life will linger
when both of us
are gone.

MURIEL RUKEYSER

(1913–80)

The Gyroscope

But this is our desire, and of its worth. . . .
Power electric-clean, gravitating outward at all points,
moving in savage fire, fusing all durable stuff
but never itself being fused with any force
homing in no hand nor breast nor sex
for buried in these lips we rise again,
bent over these plans, our faces raise to see.
Direct spears are shot outward from the conscience
fulfilling what far circuits?   Orbit of thought
what axis do you lean on, what strictnesses evade
impelled to the long curves of the will's ambition?
Centrifugal power, expanding universe
within expanding universe, what stillnesses
lie at your center resting among motion?
Study communications, looking inward, find what traffic
you may have with your silences   :   looking outward,
     survey
what you have seen of places   :
                         many times this week I seemed
to hear you speak my name
                    how you turn the flatnesses
of your cheek and will not hear my words
               Then reaching the given latitude
and longitude, we searched for the ship and found nothing
          and, gentlemen, shall we define desire
including every impulse toward psychic progress?

Roads are cut into the earth leading away from our place
at the inevitable hub.   All directions are *out*,
all desire turns outward   :   we, introspective,
continuing to find in ourselves the microcosm
imaging continents, powers, relations, reflecting
all history in a bifurcated Engine
Here is the gyroscope whirling out pulsing in tides illimitably
                    widening, live force contained
in a sphere of rigid boundary   ;   concentrate
at the locus of all forces, spinning with black speed
revolving outward perpetually, turning with its torque
all the developments of the secret will.
Flaming origins were our fathers in the heat of the earth,
pushing to the crust, water and sea-flesh,
undulant tentacles ingrown on the ocean's floor,
frondy anemones and scales' armor gave us birth.
Bring us to air, ancestors! and we breathed
the young flesh wincing against naked December.
Masters of fire, fire gave us riches, gave us life.
Masters of water, water gave us riches, gave us life,
masters of earth, earth gave us riches, gave us life.
Air mocks, and desire whirls outward in strict frenzy, leaping,
elastic circles widening from the mind,
turning constricted to the mind again.
The dynamics of desire are explained
in terms of action outward and reaction to a core
obscured and undefined, except, perhaps, as "God in
        Heaven," "God in Man,"
Elohim intermittent with the soul, recurrent
as Father and Holy Ghost, Word and responsive Word,
merging with contact in continual sunbursts,
the promise, the response, the hands laid on,
the hammer swung to the anvil, mouth fallen on mouth,
the plane nose up into an open sky.
Roads are cut, purchase is gained on our wish,
the turbines gather momentum, tools are given   :
whirl in desire, hurry to ambition, return
maintaining the soul's polarity   ;   be : fly.

DAVID PETTEYS

(1924—)

Spaces (for Samuel Beckett)

too vast
to be plotted
ever on any
conceivable
set of points,
being beyond
the scope
of Cartesian
coordinates,
the stars
and the spaces
between them
enscribe a sphere
with no discernable
circumference,
its center
simply everywhere—
even under
the table
where your
typewriter sits
in its heap
of scraps,
its cast-off
manuscripts

MARVIN COHEN
(1931—)

Statement by a Time-Dweller

Time? I don't know what it is. However, I'd be a fool not to let it *act* on me. After all, I'm *in* it, in some way.

The past? The future? They're both in my present—which moves. I have states of consciousness, in succession, one after another, always. Each moment brings a new rearrangement of parts of my past, with intentions and tendencies toward an unknown future.

If I knew what time is and how it does what it does, I'd understand myself better, and also understand an opera, a symphony, a poem, a play, a novel better.

All that moves, moves with time. I'm in it, but I can't see me in it. I see only disconnected tableaus, not vital sequences and transitions. I see no temporal structure of my mind and feelings. I'm cut off, from the true me.

I'll die not having seen. I'll die having had many not-put-together awarenesses—I'll die in my own clutter.

It makes no sense—yet, better to be than not to have been. At least I existed; but so too did everything else. It's no special distinction, to have existed. But at the least, it was *something*. And there was nothing else ever being me, but myself. It was a unique experience.

Space, and Other Places

There was no sense in walking without a purpose, so we set a destination, with the help of one of those upside-down maps that are so careful to avoid being dogmatic in pointing out any direction.

Then where did you go?

Somewhere else, at the same time.

At what point did you avoid confusion?

Equally, at none.

By then—

By then, it was all beyond recovery. Where we were, became the place to be. Anywhere else would be away from that, and left there. So, we stayed. Much later—but that's another story—we moved.

Picasso Becomes Folded Away in Space until His Disappearance Is Scattered into the Visual Round Everywhere Flat Unfolding Forms of All the Nothings

Picasso got to the point where he was unable to see anything flat anymore. So he drove to an eye specialist, in a real car, a distance of over one hundred miles across actual country—simply by visualizing it. He was at home during the whole journey—though he allowed it to take place without inconveniencing himself to get up from his chair, which was being sat continually on by his very old body on top of which were eyes that had lost the ability even to conceive of flatness. Space gave him a medal for the way he invaded it all at once, simply without bothering to shift his view. His senile lazy decline into an assault on space from all sides: and space could only be flattered for such singular attention. The man was artistically an Einstein; silhouettes collapsed at his sigh, and facades were eaten away; veneers lost their mask-like quality, and deception had to hide. Illusion turned honest, for once, and was arrested on the spot. Picasso will die in a flat box. The eyes have jumped away already, and go between things: like flies in and around everything. Visuality has had her virginity penetrated—and her roaming bridegroom still has his bachelor cunning. Sight blushes, for her shame. The rape is on permanent view. The stages slip by slowly; this performing exhibitionist is Art; life retreats to a safe remove while the spectacle keeps on happening, with dice-like combinations from eyes endless in their motion, firmly shifting at every turn, yielding a corridor behind each tableau, and the corridor renews itself, emptying time of that human parasite: Impatience.

SIV CEDERING
(1939—)

From *Letters from the Astronomers*

V. ALBERT EINSTEIN (1879–1955)

Yes, I have written
the President. I have told him
that if there is a nuclear war, the Fourth
World War will be fought
with sticks and stones.

Words do hurt me,
and there is no change in my heart
condition, but I am trying to complete
my unified field
theory. I cannot believe
that God is playing roulette
with the world. The mystery
must be locked up
in the elemental infrastructures.

Forgive me for using
scrap paper. The other day
when my wife and I

were being shown
the huge reflecting telescope
at Mount Wilson observatory,
she asked why the instruments
were so large.

On being told
that they were trying to discern
the shape and makeup of the whole
universe, she said:
My husband does that
on the back
of an old
envelope.

RICHARD EBERHART
(1904—)

On Shooting Particles beyond the World

"White Sands, N.M., Dec. 18 (U.P.) 'We first throw a little something into the skies,' Zwicky said. 'Then a little more, then a shipload of instruments—then ourselves.'"

On this day man's disgust is known
Incipient before but now full blown
With minor wars of major consequence,
Duly building empirical delusions.

Now this little creature in a rage
Like new-born infant screaming compleat angler

Objects to the whole globe itself
And with a vicious lunge he throws

Metal particles beyond the orbit of mankind.
Beethoven shaking his fist at death,
A giant dignity in human terms,
Is nothing to this imbecile metal fury.

The world is too much for him. The green
Of earth is not enough, love's deities,
Peaceful intercourse, happiness of nations,
The wild animal dazzled on the desert.

If the maniac would only realize
The comforts of his padded cell
He would have penetrated the
Impenetrability of the spiritual.

It is not intelligent to go too far.
How he frets that he can't go too!
But his particles would maim a star,
His free-floating bombards rock the room.

Good Boy! We pat the baby to eructate,
We pat him then for eructation.
Good Boy Man! Your innards are put out,
From now all space will be your vomitorium.

The atom bomb accepted this world,
Its hatred of man blew death in his face.
But not content, he'll send slugs beyond,
His particles of intellect will spit on the sun.

Not God he'll catch, in the mystery of space.
He flaunts his own out-cast state
As he throws his imperfections outward bound,
And his shout that gives a hissing sound.

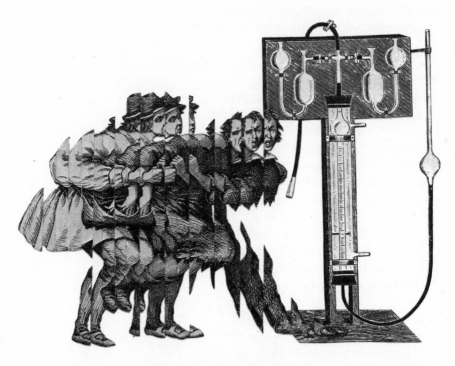

JOSEPHINE MILES
(1911—)

Physics

The mean life of a free neutron, does it exist
In its own moving frame a quarter-hour?
In decibel, gram, ohm, slug, volt, watt, does it exist?
Not objects answer us, not the hand or eye,
But particles out of sight.

Does a book in equilibrium on a shelf
Compose its powers? It upsets my mind.
Turbulent flow, function of force times distance,

The sledge with a steel head
Is energy transformed.

Illusion boils a water into cold,
A speed of pulse slows by chronometer,
A camera's iris diaphragm opens wide
To faint light. Lenses
Render to us figures equivocal.

Sight in its vacuum, sound in its medium strike so aslant
The thunder relishes a laggard roll,
And as long waves of low-pitched sounds bend around
    corners,
Building-corners cut off a high wrought bell,
To set its nodes and loops vibrating symmetrically at the
    surface.

Neutron transformed, neutron become again,
In glass and silk, tracing a straight world line,
Exists in its gravitational electromagnetic fields
Trembling, though beyond sight,
Initially at rest.

CHEMISTRY

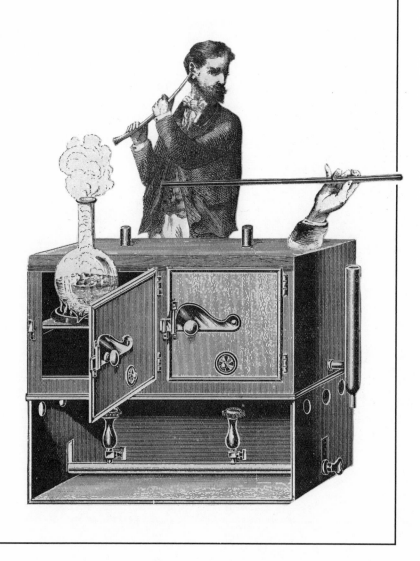

INTRODUCTION

> Nobody, I suppose, could devote many years to the
> study of chemical kinetics without being deeply con-
> scious of the fascination of time and change: this is
> something that goes outside science into poetry; but
> science, subject to the rigid necessity of always seek-
> ing closer approximations to the truth, itself contains
> many poetical elements.
> —Cyril N. Hinshelwood, Nobel lecture, 1956

The origins of the word *chemistry* reveal its close ties to the
ancient crafts. The ancient name for Eygpt was *Khem,* and the
Arabs called the art of transmuting things by fire *Al Kemia,*
"the Egyptian art." *Al Kemia* became *alchemy* and later *chemistry*. As this derivation suggests, ancient Egypt was famous
for its crafts, especially techniques for glazing ceramics and
metallurgy, which developed so that the Egyptians could produce thousands of bronze images of gods for offerings in
temples.

Obviously, for the early Egyptian craftsmen practical considerations were foremost. How to make the glaze blue or
how to make copper tools harder were their typical concerns.
It is just as clear that the Greek civilization was the first to
attempt a more theoretical knowledge of how the world is
being held together. Plato, in the *Timaeus,* suggests that the
basic building blocks of the universe are the four traditional
elements: earth, air, fire, and water. (He proposes that there is
also a fifth element, quintessence, but never expands on this.)
Each of the four elements is correlated with one of the regular
solids: earth to the cube, fire to the tetrahedron, air to the
octahedron, and water to the icosahedron. Since there is also

a dodecahedron, this is assigned to the shadowy fifth element. In this theory everything that exists is regarded as a combination of these elements. A modified, elemental approach to matter is still basic to modern chemistry.

Plato's ideas had a profound influence on those who followed him. Pliny merely echoed the *Timaeus* in his discussions of the elements. The school of Epicurus of Samos presented atoms as the basic building blocks, and the Roman poet Lucretius reasserted this theory two centuries later. His epic poem *"De rerum natura"* ("On the Nature of Things") hinged on the notion of infinite, random combinations of matter. He used this notion as an argument for the plurality of worlds and against the fear of death. Like many of the ancients, he gave his interest in matter psychological ramifications.

Indeed, for most of its history chemistry has been associated with magic and power, the secrets and control of the world's elemental stuff. Given this background, it is not surprising that alchemy developed in Alexandria, a Greek city in Egypt. Here the philosophers with their theoretical concerns could observe the craftsmen with their empirical concerns. From this union arose alchemy, an attempt to bring about specific empirical results based on the manipulation of elements but guided by theoretical principles.

Of all the crafts, metallurgy is probably most closely allied with alchemy. To a great extent both alchemy and metallurgy depend on fire. (Fire is the only element not inhabited by animals, according to ancient theory.) As early as ten thousand years ago it was realized that naturally occurring copper could be hammered and worked into tools. But it wasn't until four thousand years later that man learned to release copper from ore through the use of fire. Still, the pure metal had certain weaknesses. Like all metals, it is made of thin crystalline layers that slide when pulled. Thus, under stress, copper is inclined to tear. The ancients reasoned that if something could be introduced into the copper that would make it less regular and slippery, then it would be stronger.

The discovery that copper combined with 5 to 25 percent tin became a strong alloy—bronze—must have seemed like a miracle to the early craftsmen. For although tin is softer than copper, when combined with copper it forms a material stronger than either. On the same principle, it was discovered that fire could release iron from its ore and that the relatively soft iron combined with scant carbon (less than 5 percent) produced hard steel. It is easy to see how the earliest chemistry became linked to power of one kind or another.

While the craftsmen were concerned with making metals stronger or more flexible, the alchemists were concerned with a different application of the metallurgist's fire. It would be an oversimplification to say that the prime task of the alchemist was to turn base metal into gold. While this has become the stereotype of the alchemic goal, the true story is far more complex.

The alchemist worked on several levels at once. Each physical element had for him a spiritual property. Thus gold, which didn't tarnish, represented purity, while mercury, because of its protean shape shifting, represented transformation. Each of the elements, moreover, was associated with a planet, so that the complete alchemic endeavor was a physical-spiritual-astrological drama. The ultimate hope of the alchemist was to undergo his own spiritual transformation from impure to incorruptible, just as his base metals turned to gold. Like the elements in his vessels, the alchemist would endure trial by fire to obtain purity. The language used by alchemists indicates the various levels of the operations. To transmute the base metal it was necessary to produce "the tincture of the soul." This tincture was believed to be mercury, which had the power to make the sick (base) metal well (precious).

In the medieval poem *The Romance of the Rose* Jean de Meun considers that alchemy is properly conceived of as an applied art akin to metallurgy and glassmaking. The poem, in part an allegory of courtly love, is also a compendium of current philosophy and science. It is filled with reverence for "Nature's subtlety" which, the poet argues, can be deciphered by the true artist—the alchemist—though not the sophist!

Like Jean de Meun, many later writers were fascinated by mysterious chemical transformations apparent in nature. Francis Bacon interpreted these mysteries as an allegory that identified matter with Proteus, the god of change. His fable seems a prophetic assertion of the first law of thermodynamics, that matter can neither be created nor destroyed—except, Bacon argues, by the Creator himself. Bacon, who likes to emphasize the positive connection between knowledge and power, nevertheless regards the full nature of matter as something beyond the limited power of man.

Other religious poets of the seventeenth century discourage alchemy more overtly. For John Donne it is pure "imposture," which he turns into a pessimistic equation between the search for an elixir and the search for "minde in women." He finds both tinged with a malodorous reality instead of an angel's breath.

Among the many misconceptions revolving around alchemy was that alchemists were charlatans who duped the ignorant public with cheap tricks like gold-plating lead—or worse. In his comedy *The Alchemist* Ben Jonson invents a multitude of deceptions that pass for science among avaricious people drawn to alchemy by their own greed. Jonson's characters have themselves an elemental quality. They are called "humour" figures based on contemporary medicine and psychology that typed the human personality according to its proportion of body fluids. Thus were men reduced to the old four elements by their balance of blood (hot and moist), phlegm (cold and moist), yellow bile (hot and dry), and black bile (cold and dry). It is not surprising that a dramatist who could animate such a theory as this should find alchemy a natural subject. Jonson's alchemist, called Subtle, knows well, as his name suggests, how to read the disposition of the greedy. Likewise Face, his manservant, has a mask for every client. In the scenes excerpted here, the pair meet with Epicure Mammon, a knight, and Surly, a gamester, who hang with passionate self-interest on the progress of a transmutation.

In Jonson's age there undoubtedly were tricksters out to bilk greedy investors, but these were not the true alchemists. From the notebooks left by the practitioners themselves, it is clear that alchemists were misguided but honest men. Those such as Paracelsus (1493–1541), the father of modern pharmacology, and John Dee, Elizabeth I's astrologer, never claimed to have discovered the philosopher's stone or to have transmuted lead into gold. They were attempting it, often believed they were close to it, but never claimed more than they had actually accomplished.

Another common misconception about alchemy is just how much modern chemistry owes to this maligned ancestor. While the alchemists did develop a scientific method and did keep careful records of their experiments, to some extent they hindered the progress of chemistry by the misleading theoretical scaffolding that they built around their empirical findings.

This was the lesson learned by the young Victor Frankenstein in Mary Shelley's novel. Interestingly, her hero may very well have owed something to the character of her husband, the poet Percy Bysshe Shelley. In one of his poems, "Alastor: or, The Spirit of Solitude," he writes about a longing for "Medea's wondrous alchemy," and in his own life he was very much the amateur scientist. Shelley, like many men of his century, was particularly fascinated with the connection between electricity and chemistry that was so dramatically revealed by the electrical muscle experiments of Luigi Galvani (1737–1798).

Like the poet, Victor Frankenstein is a scientific enthusiast who longs for power over life and death. First he trains on alchemic texts but to no avail. Chemistry, he is told, in this "enlightened and scientific age" holds the real key. With this information he proceeds in the right direction to his astonishing experiment, which sparks a monster into being. The novel itself is subtitled "A Modern Prometheus," after the mythic god who gave the gift of fire to man. In Shelley's age electricity seemed to be a modern version of that gift and was deeply connected with the pursuit of chemistry and the quest for power.

Mary Shelley's gothic novel is also quite correct in its suggestion that modern chemistry probably owes more to medicine and mathematics than to alchemy. As it happens, Dr. Frankenstein's early dabbling in Paracelsus was not so terribly misguided in this respect, for Paracelsus was primarily a physician and then an alchemist. He discovered that certain substances when ingested by a patient could have marked effects on his health. In addition, he saw that it was important to measure the specific quantities of the substances administered. Thus, from his procedure the analysis of chemicals became quantitative.

Indeed, from the sixteenth century forward a host of other nonmedicinal experiments based on measurement advanced chemistry as a science. Mining was becoming a critical industry in Europe, and books on techniques for separating and testing elements began to appear. From this flurry of interest one technological advance appeared that is still central to modern chemistry—distillation. Using this method of combined heating and cooling, the "essence" of a substance could be obtained. Elaborate stills were constructed and new elements discovered by improved distillation techniques. It became clear that if one were to describe the building blocks of the universe, the original four elements—earth, air, fire, and water—were totally unsatisfactory. In 1661 Robert Boyle published *The Sceptical Chymist,* affirming this position and paving the way for the era of modern chemistry.

By the late seventeenth century there was also a growing belief in "better things for better living through chemistry"— to borrow a more recent motto. Air pumps, pistons, and Leyden jars made for parlor conversation, and men of no specific training gave themselves to scientific inquiry that led to the Industrial Revolution. Quick to see the connection between chemistry and industry, Erasmus Darwin (the grandfather of Charles) wrote long, impassioned verses on the potential energy concealed in elemental resources. He extolled the power of iron, clay, and not least of all lightning. Darwin's rhapsody on Benjamin Franklin's electrical experiments as well as Franklin's own experimental writings illustrate no less than *Frankenstein* the romantic aspirations of chemistry at a crucial point in time: the great age of political and industrial revolution.

A political revolutionary, Joseph Priestly (1733–1804), was the amateur chemist who truly began the new age of modern chemistry. His specialty was the study of gases, and he discovered that air was composed of various invisible gases that were separate and distinct elements. Before Priestly, the different properties of gases had been observed but they had been considered as one element—air—with different impurities. Priestly isolated and discovered the properties of nitric oxide as well as oxygen, which he called "dephlogisticated air." He saw that candles burned brightly in it and mice breathed easily in it. Along with his other credits is the invention of soda water which, when it was discovered to have a pleasing taste and efflorescence, brought him considerable publicity.

Priestly was most interested in the qualities rather than the quantities of the gases he studied. He had little patience with the balance scale, and his notes are unreliable with regard to quantity, particularly when compared to those of his contemporary Antoine-Laurent Lavoisier (1743–94), who ushered in the use of quantitative analysis with an experiment duplicating a basic alchemical exercise in transformation that was commonly done in the Middle Ages. First cinnabar (sulfide of mercury) was heated. The heat drove off the sulfur and left pure mercury. When this was heated again, as if by magic the cinnabar seemed to reappear. This was not actually the case. When heated and combined with oxygen, mercury yields an oxide of mercury that is red and looks like cinnabar. Lavoisier did the experiment but measured the quantities exchanged. He burned mercury and noted the precise amount of oxygen that was taken up from a closed vessel. Then he burned the mercuric oxide to drive off the oxygen and discovered that the amount of oxygen expelled was equal to the amount taken up. Lavoisier's demonstration that the two elements can combine and can be separated in quantifiable ways heralded the quantification of chemistry and overthrew once and for all alchemy's pseudo-transformations.

In retrospect, it is difficult to grasp how deeply indebted early modern chemistry was to home-laboratory experiments undertaken by inquisitive people with a simple sense of curiosity. Occasionally, these people included poets, like Shelley and Samuel Taylor Coleridge, who mingled among scientific men because chemistry stirred their imaginations. Coleridge, who was notoriously addicted to laudanum, was particularly enthusiastic about research on nitrous oxide (laughing gas) undertaken by his chemist-friend Sir Humphry Davy. Like Lavoisier's work with oxygen, Davy's study of nitrous oxide was related to the current interest in gases and respiration; it was also, like Lavoisier's work, dependent on quantitative measurement. But Davy made another connection between chemistry and philosophy, and his research on laughing gas was particularly conducive to a qualitative study on the exhilarating sensations expressed by romantic poetry. Coleridge certainly was the ideal experimental subject since he could describe the effects of the gas with a poet's sensitivity.

While Davy quantified sensational effects, English scientist John Dalton (1766–1844) preferred to concern himself more soberly with causes. He wanted to know why it is that oxygen and hydrogen always combine in the same proportions to form water. His questioning arose from a simple observation that required an explanation. When gases of different weights are introduced into a container, they do not do what we expect— form layers with the heaviest on the bottom and the lightest on top. Rather, they diffuse and form a homogenous gas.

Dalton's attempt to explain the rules governing the combination of gases led to his theory of atomism. He realized that the Greek atomic theory was correct and suggested that "ultimate elementary particles," or atoms, were the building blocks of the universe and that they might have different weights. Consider carbon dioxide (CO_2) and water H_2O). The weight of oxygen (O_2) that will produce one unit of CO_2 will produce two units of H_2O. If we take away the two oxygen atoms from the CO_2 and from the two units of water, we have precisely the right proportion of H and C for methane (CH_4). The weight is constant. Dalton worked out atomic weights for sixteen elements and saw the start of modern atomic theory.

Dalton himself was occasionally metaphoric in his attempt to describe the behavior of microcosmic ultimate particles that could not be seen by the naked eye. He likened them to stars and planets held in perfect balance in the universe, and this pictorial view of atomic structure filtered almost immediately into popular thought. Even the American poet Emily Dickinson, generally regarded as reclusive and conservative, was conscious of atoms as little solar systems of herself. The "Chemical Conviction" of a closed system in which matter is not lost gives her some cause to argue against fears of her own disintegration. The connection she makes between chemical life force and electricity is, moreover, true to the spirit of nineteenth-century chemistry.

In the fifty years after Dalton new elements continued to be discovered, and it was found that electricity could break down compounds into more basic constituents. In addition, as the number of identifiable elements increased, it became apparent that there were similarities or family likenesses. According to Dalton, what distinguished each element from the other was its atomic weight. But how could this single property cause the differences and similarities?

The man who attempted to answer this question was Dmitry Ivanovich Mendeleev (1834–1907). He wrote each element, with its atomic weight and properties, on an index card. Then he spread the cards out in order of atomic weight and found that they fell into families. When he placed them in vertical columns of seven he found that reading across (horizontally) there were family resemblances.

hydrogen^1	lithium^7	sodium^{23}	potassium^{39}
	beryllium^9	magnesium^{24}	calcium^{40}
	boron^{11}	aluminum^{27}	
	carbon^{12}	silicon^{28}	titanium^{48}
	nitrogen^{14}	phosphorus^{31}	
	oxygen^{16}	sulphur^{32}	
	fluorine^{19}	chlorine^{35}	bromine^{80}

Hydrogen made the pattern go out of line so he wisely put it off to the side. Looking at this pattern, he could see which elements didn't fit and would leave gaps. For example, titanium could not be grouped with boron and aluminum, so he left a gap and placed titanium with carbon and silicon. Although there were gaps in the table, Mendeleev could predict the properties of the missing elements. (He knew only about sixty of the ninety-two elements.) He could tell how much its oxide would weigh and if it would be a gas or a solid. Furthermore, he made these predictions with great accuracy, despite the discrepancies that he never solved.

The reduction of the universe to ultimate, unseen particles both fascinated and disturbed the nineteenth-century imagination. To what does man reduce? is a characteristically Victorian question. Matthew Arnold found that brooding on the elements took him back to pre-Socratic Greek thought and the problem of whether mind could be distinguished from essential particles. His long poem "Empedocles on Etna" is a meditation as much about the reductive values of his own age as it is about the past.

In chemistry as in biology and the other sciences the nineteenth-century theorists began to investigate the unseen with mixed confidence and trepidation. Mendeleyev never did answer the question of why there were family resemblances among the elements; he merely organized and described them. The answer was supplied in 1897 by Sir Joseph John Thomson (1856–1940) of Cambridge: The atom was not indivisible it merely hid a more complex structure. Thomson discovered the electron and determined that the properties of each element depend to a great extent on the number of electrons in its atom. He showed that the number of electrons corresponds to the place it has in the periodic chart, and this he called atomic number. This discovery enabled him to rearrange the periodic table of elements as first presented by Mendeleyev. With the focus shifted from atomic weight to atomic number, chemistry joined modern physics in becoming a structural science.

This new age is perhaps best symbolized by the experiments of Max von Laue (1879–1960) and William Henry Bragg (1862–1942). Their work on the X-ray diffraction of crystals inspired two very fine literary pieces by Italo Calvino and Loren Eiseley. Calvino's short story "Crystals," from the collection *t zero*, places his cosmic hero, Qfwfq, and his girl friend, Vug, in the crystal world of earth's incandescence, which is at once primordial and New York City. Calvino's vision and wit bridge the gap between theory and human emotion. Eiseley, too, responds to the twists and distortions of crystalline form, which correspond to the secret individuality of the mind.

In all respects the twentieth century has been preoccupied with the invisible structures of reality. The chemistry of electron bonding emphasizes the unseen vitality of seemingly inert substances. And for this reason the chemical bond has been taken as a metaphor for mind and personality. John Updike, truly an Erasmus Darwin of our age, has written a chemical allegory of the human self in which he discovers that its solidity proves to be giddy with intricate flaws and movements.

PLATO

(428–348 B.C.)

From *Timaeus*

The various elements had different places before they were arranged so as to form the universe. At first, they were all without reason and measure. But when the world began to get into order, fire and water and earth and air had only certain faint traces of themselves, and were altogether such as everything might be expected to be in the absence of God; this, I say, was their nature at that time, and God fashioned them by form and number. Let it be consistently maintained by us in all that we say that God made them as far as possible the fairest and best, out of things which were not fair and good. And now I will endeavour to show you the disposition and generation of them by an unaccustomed argument, which I am compelled to use; but I believe that you will be able to follow me, for your education has made you familiar with the methods of science.

In the first place, then, as is evident to all, fire and earth and water and air are bodies. And every sort of body possesses solidity, and every solid must necessarily be contained in planes; and every plane rectilinear figure is composed of triangles; and all triangles are originally of two kinds, both of which are made up of one right and two acute angles; one of them has at either end of the base the half of a divided right angle, having equal sides, while in the other the right angle is divided into unequal parts, having unequal sides. These, then, proceeding by a combination of probability with demonstration, we assume to be the original elements of fire and the other bodies; but the principles which are prior to these God only knows, and he of men who is the friend of God. And next we have to determine what are the four most beautiful bodies which are unlike one another, and of which some are capable of resolution into one another; for having discovered thus much, we shall know the true origin of earth and fire and of the proportionate and intermediate elements. And then we shall not be willing to allow that there are any distinct kinds of visible bodies fairer than these. Wherefore we must endeavour to construct the four forms of bodies which excel in beauty, and then we shall be able to say that we have sufficiently apprehended their nature. Now of the two triangles, the isosceles has one form only; the scalene or unequal-sided has an infinite number. Of the infinite forms we must select the most beautiful, if we are to proceed in due order, and any one who can point out a more beautiful form than ours for the construction of these bodies, shall carry off the palm, not as an enemy, but as a friend. Now, the one which we maintain to be the most beautiful of all the many triangles (and we need not speak of the others) is that of which the double forms a third triangle which is equilateral; the reason of this would be long to tell; he who disproves what we are saying, and shows that we are mistaken, may claim a friendly victory. Then let us choose two triangles, out of which fire and the other elements have been constructed, one isosceles, the other having the square of the longer side equal to three times the square of the lesser side.

Now is the time to explain what was before obscurely said:

there was an error in imagining that all the four elements might be generated by and into one another; this, I say, was an erroneous supposition, for there are generated from the triangles which we have selected four kinds—three from the one which has the sides unequal; the fourth alone is framed out of the isosceles triangle. Hence they cannot all be resolved into one another, a great number of small bodies being combined into a few large ones, or the converse. But three of them can be thus resolved and compounded, for they all spring from one, and when the greater bodies are broken up, many small bodies will spring up out of them and take their own proper figures; or, again, when many small bodies are dissolved into their triangles, if they become one, they will form one large mass of another kind. So much for their passage into one another. I have now to speak of their several kinds, and show out of what combinations of numbers each of them was formed. The first will be the simplest and smallest construction, and its element is that triangle which has its hypotenuse twice the lesser side. When two such triangles are joined at the diagonal, and this is repeated three times, and the triangles rest their diagonals and shorter sides on the same point as a centre, a single equilateral triangle is formed out of six triangles; and four equilateral triangles, if put together, make out of every three plane angles one solid angle, being that which is nearest to the most obtuse of plane angles; and out of the combination of these four angles arises the first solid form which distributes into equal and similar parts the whole circle in which it is inscribed. The second species of solid is formed out of the same triangles, which unite as eight equilateral triangles and form one solid angle out of four plane angles, and out of six such angles the second body is completed. And the third body is made up of 120 triangular elements, forming twelve solid angles, each of them included in five plane equilateral triangles, having altogether twenty bases, each of which is an equilateral triangle. The one element [that is, the triangle which has its hypotenuse twice the lesser side] having generated these figures, generated no more; but the isosceles triangle produced the fourth elementary figure, which is compounded of four such triangles, joining their right angles in a centre, and forming one equilateral quadrangle. Six of these united form eight solid angles, each of which is made by the combination of three plane right angles; the figure of the body thus composed is a cube, having six plane quadrangular equilateral bases. There was yet a fifth combination which God used in the delineation of the universe.

Now, he who, duly reflecting on all this, enquires whether the worlds are to be regarded as indefinite or definite in number, will be of opinion that the notion of their indefiniteness is characteristic of a sadly indefinite and ignorant mind. He, however, who raises the question whether they are to be truly regarded as one or five, takes up a more reasonable position. Arguing from probabilities, I am of opinion that they are one; another, regarding the question from another point of view, will be of another mind. But, leaving this enquiry, let us proceed to distribute the elementary forms, which have now been created in idea, among the four elements.

To earth, then, let us assign the cubical form; for earth is the most immoveable of the four and the most plastic of all bodies, and that which has the most stable bases must of necessity be of such a nature. Now, of the triangles which we assumed at first, that which has two equal sides is by nature more firmly based than that which has unequal sides; and of the compound figures which are formed out of either, the plane equilateral quadrangle has necessarily, a more stable basis than the equilateral triangle, both in the whole and in the parts. Wherefore, in assigning this figure to earth, we adhere to probability; and to water we assign that one of the remaining forms which is the least moveable; and the most moveable of them to fire; and to air that which is intermediate. Also we assign the smallest body to fire, and the greatest to water, and the intermediate in size to air; and, again, the acutest body to fire, and the next in acuteness to air, and the third to water. Of all these elements, that which has the fewest bases must necessarily be the most moveable, for it must be the acutest and most penetrating in every way, and also the lightest as being composed of the smallest number of similar particles: and the second body has similar properties in a second degree, and the third body in the third degree. Let it be agreed, then, both according to strict reason and according to probability, that the pyramid is the solid which is the original element and seed of fire; and let us assign the element which was next in the order of generation to air, and the third to water. We must imagine all these to be so small that no single particle of any of the four kinds is seen by us on account of their smallness: but when many of them are collected together their aggregates are seen. And the ratios of their numbers, motions, and other properties, everywhere God, as far as necessity allowed or gave consent, has exactly perfected, and harmonised in due proportion.

From all that we have just been saying about the elements or kinds, the most probable conclusion is as follows:—earth, when meeting with fire and dissolved by its sharpness, whether the dissolution take place in the fire itself or perhaps in some mass of air or water, is borne hither and thither, until its parts, meeting together and mutually harmonising, again become earth; for they can never take any other form. But water, when divided by fire or by air, on re-forming, may become one part fire and two parts air; and a single volume of air divided becomes two of fire. Again, when a small body of fire is contained in a larger body of air or water or earth, and both are moving, and the fire struggling is overcome and broken up, then two volumes of fire form one volume of air; and when air is overcome and cut up into small pieces, two and a half parts of air are condensed into one part of water. Let us consider the matter in another way. When one of the other elements is fastened upon by fire, and is cut by the sharpness of its angles and sides, it coalesces with the fire, and then ceases to be cut by them any longer. For no element which is one and the same with itself can be changed by or change another of the same kind and in the same state. But so long as in the process of transition the weaker is fighting against the stronger, the dissolution continues. Again, when a few small particles, enclosed in many larger ones, are in process of decomposition and extinction, they only cease from their tendency to extinction when they consent to pass into the conquering nature, and fire becomes air and air water. But if bodies of another kind go and attack them [i.e., the small particles], the latter continue to be dissolved until, being completely forced back and dispersed, they make their escape to their own kindred, or else, being overcome and assimilated to the conquering power, they remain where they are and dwell with their victors, and from being many become one. And owing to these affections, all things are changing their place, for by the motion of the receiving vessel the bulk of each class is distributed into its proper place; but those things which become unlike themselves and like other things, are hurried by the shaking into the place of the things to which they grow like.

PLINY THE ELDER

(23–79)

From *Natural History*

BOOK II

Chapter 4—Of the Elements and the Planets

I do not find that any one has doubted that there are four elements. The highest of these is supposed to be fire, and hence proceed the eyes of so many glittering stars. The next is that spirit, which both the Greeks and ourselves call by the same name, air. It is by the force of this vital principle, pervading all things and mingling with all, that the earth, together with the fourth element, water, is balanced in the middle of space. These are mutually bound together, the lighter being restrained by the heavier, so that they cannot fly off; while, on the contrary, from the lighter tending upwards, the heavier are so suspended, that they cannot fall down. Thus, by an equal tendency in an opposite direction, each of them remains in its appropriate place, bound together by the never-ceasing revolution of the world, which always turning on itself, the earth falls to the lowest part and is in the middle of the whole, while it remains suspended in the centre, and, as it were, balancing this centre, in which it is suspended. So that it alone remains immoveable, whilst all things revolve round it, being connected with every other part, whilst they all rest upon it.

Between this body and the heavens there are suspended, in this aërial spirit, seven stars, separated by determinate spaces, which, on account of their motion, we call wandering, although, in reality, none are less so. The sun is carried along in the midst of these, a body of great size and power, the ruler, not only of the seasons and of the different climates, but also of the stars themselves and of the heavens. When we consider his operations, we must regard him as the life, or rather the mind of the universe, the chief regulator and the God of nature; he also lends his light to the other stars. He is most illustrious and excellent, beholding all things and hearing all things, which, I perceive, is ascribed to him exclusively by the prince of poets, Homer.

LUCRETIUS

[TITUS LUCRETIUS CARUS]

(96–55 B.C.)

From *De rerum natura*

BOOK II

But now I'll sing, do you attend, how seed
Proceeds to make, and to dissolve things made:
What drives them forward to their tedious race,

What makes them run thro' all the mighty space.
  'Tis certain now no seed to seed adheres
Unmov'd, and fix'd: for ev'ry thing appears
Worn out, and wasted by devouring years;
Still wasting, still it vanishes away,
And yet the mass of things feels no decay:
For when those bodies part, the things grow less,
And old: but they do flourish, and encrease,
To which they join; hence too they fly away;
So things by turns increase, by turns decay:
Like racers, bear the lamp of life, and live,
And their race done, their lamp to others give.
And so the mass renews: few years deface
One kind, and straight another takes the place.
  But if you think the seeds can rest and make
A change by rest; how great is the mistake?
For since they thro' the boundless vacuum rove,
By their own weight, or other's stroke they move,
For when they meet and strike, that furious play
Makes each of them reflect a diff'rent way:
For both are perfect solids, and nought lies
Behind, to stop their motion as they rise.
  But that you may conceive how thus they move;
Consider, that my former reasons prove
That seeds seek not the midst, and that the space
Is infinite, and knows no lowest place;
And therefore seeds can never end their race:
But always move, and in a various round.
  Some, when they meet, and rudely strike, rebound
To a great distance; others, when they jar,
Will part too, and rebound, but not so far:
Now these small seeds, that are more closely join'd,
And tremble, in a little space confin'd,
Stopt by their mutual twinings, stones compose
Iron, or steel, or bodies like to those;
But those, that swim in a wide void alone,
And make their quick and large rebounds, or run
Thro' a large space, compose the air, and sun.
  Besides these two, there is another kind;
Bodies from union free, and unconfin'd;

With others ne'er in friendly motions join'd.
Of these there's a familiar instance ———
For look where'er the glitt'ring sunbeams come
Thro' narrow chinks, into a darken'd room;
A thousand little bodies straight appear
In the small streams of light, and wander there:
For ever fight, reject all shews of peace;
Now meet, now part again, and never cease:
Hence we may judge how th' atoms always strove
Thro' the vast empty space, and how they move.
Such knowledge from mean instances we get,
And easily from small things rise to great.
  But mark this instance well, and learn from thence
What motions vex the seeds, tho hid from sense:
For here you may behold, by secret blows
How bodies turn'd, their line of motions lose:
How beaten backward, and with wanton play,
Now this, now that, and ev'ry other way.
All have their motions from their seeds; for those
Move of themselves, and then with secret blows
Strike on the small molecule; they receive
The swift impression, and to greater give;
Thus they begin from the first seeds; and thence
Go on by just degrees, and move our sense.
For look within the little beam of light
You see them strike; but what blow makes them fight
Is undiscern'd, and hidden from our sight.
  And yet how swift the atoms motions are,
This foll'wing instance will in short declare:
For when the morning climbs the eastern skies,
And tuneful birds salute her early rise;
In ev'ry grove and wood with joy appear,
And fill with rav'shing sounds the yielding air:
How swift the beams of the bright rising sun
Shoot forth! Their race is finish'd when begun:
From heav'n to earth they take their hasty flight,
And guild the distant globe with gawdy light.
But this thin vapour, and this glitt'ring ray,
Thro' a meer void, make not their easy way;
But with much trouble force a passage thro'

Resisting air; and therefore move more slow;
Nor are they seeds, but little bodies join'd;
And adverse motions in small space confin'd:
And therefore from without resisting force,
And inbred jars must stop their eager course;
But solid seeds, that move thro' empty space;
Whom nothing from without resists; than light
And beams more swift, must make their hasty flight;
And in that time a larger distance fly,
While the sun's lazy beams creep thro' our sky:
For they by counsel can not move more slow;
Or stop to make inquiry, or to know
How they must work, on what design they go.
    But some, dull souls! Think matter can not move
Into fit shapes, without the pow'rs above:
Nor make the various seasons of the year
So fit for man; nor fruit, nor bushes bear,
Nor other things, which pleasures prompts, could do:
Pleasure, that guide of life, and mistress too!
That we should seek love's generous embrace,
And thence renew frail man's decaying race:
And therefore fancy that the gods did make
And rule this all. How great is that mistake!
          [Translated by Thomas Creech]

JEAN DE MEUN

(?–1305)

From *The Romance of the Rose*

Though Art should learn so much of Alchemy as to tint all the
metals with color, he would die sooner than transmute species, his work being at best to return them to their primitive
matter; though he worked as long as he lived, never would he

equal Nature. And if he wished to undertake the reduction to prime matter, he would have to know how to obtain, when he makes his elixir, the determining ratio from which the new form emerges, the proportion which distinguishes substances among themselves by those specific differences that define their essential natures.

However, it is well known that Alchemy is a true art; whoever practices it wisely will make miracles, for whatever the case may be concerning species, particular bodies subjected to intelligent preparations are mutable in many ways that enable them to shift their natures among themselves by diverse elaborations so that their alterations entitle them to enter into different categories. Cannot one see how the master glassmakers transform ferns to ashes first and then to glass? And yet glass is not fern and fern not glass. When thunder rumbles and lightning flashes, often one observes falling hailstones, stones of vapor that are not stone at all. The sage can tell the causes of such material alterations. They are species of transmutation in which individuals deviate in substance and shape—in the former case by the intervention of Art and in the latter by the hand of Nature.

Thus, one would be able to make metals if one could succeed in liberating the pure ore from the ordure that contaminates it; by their like complexions these cousins are drawn together, for they are of the same essential matter. Such is the disposition of their elements. The philosophy books in effect tell us that diverse metals are born in mines of sulfur and quicksilver (mercury). Whoever would be skillful in preparing these spirits in such a way that they might have the virtue necessary to enter into bodies and fix themselves there until they find them purified and the sulfur—white or red—past burning, might then have true mastery of metals. For the master alchemists can bring to birth pure gold from silver, by adding to it weight and color from ingredients less costly; and from fine gold they then make precious stones, transparent and truly remarkable. The masters also plunder from other species good metals which they transmute to silver by the use of white drugs that are refined and penetrating. But sophisters can do nothing of this kind; work as much as they would, they will never equal Nature.

FRANCIS BACON

(1561-1626)

From *Wisdom of the Ancients*

XIII--PROTEUS, OR MATTER
EXPLAINED OF MATTER AND ITS CHANGES

Proteus, according to the poets, was Neptune's herdsman; an old man, and a most extraordinary prophet, who understood things past and present, as well as future; so that besides the business of divination he was the revealer and interpreter of all antiquity, and secrets of every kind. He lived in a vast cave, where his custom was to tell over his herd of seacalves at noon, and then to sleep. Whoever consulted him, had no other way of obtaining an answer, but by binding him with manacles and fetters; when he, endeavoring to free himself, would change into all kinds of shapes and miraculous forms; as of fire, water, wild beasts, etc.; till at length he resumed his own shape again.

Explanation.—This fable seems to point at the secrets of nature, and the states of matter. For the person of Proteus denotes matter, the oldest of all things, after God himself; that resides, as in a cave, under the vast concavity of the heavens. He is represented as the servant of Neptune, because the various operations and modifications of matter are principally wrought in a fluid state. The herd, or flock of Proteus, seems to be no other than the several kinds of animals, plants, and minerals, in which matter appears to diffuse and spend itself; so that after having formed these several species, and as it were finished its task, it seems to sleep and repose, without otherwise attempting to produce any new ones. And this is the moral of Proteus' counting his herd, then going to sleep.

This is said to be done at noon, not in the morning or evening; by which is meant the time best fitted and disposed for the production of species, from a matter duly prepared, and made ready beforehand, and now lying in a middle state, between its first rudiments and decline; which, we learn from sacred history, was the case at the time of the creation; when, by the efficacy of the divine command, matter directly came together, without any transformation or intermediate changes, which it affects; instantly obeyed the order, and appeared in the form of creatures.

And thus far the fable teaches of Proteus, and his flock, at liberty and unrestrained. For the universe, with the common structures and fabrics of the creatures, is the face of matter, not under constraint, or as the flock wrought upon and tortured by human means. But if any skillful minister of nature shall apply force to matter, and by design torture and vex it, in order to [effect] its annihilation, it, on the contrary, being brought under this necessity, changes and transforms itself into a strange variety of shapes and appearances; for nothing but the power of the Creator can annihilate, or truly destroy it; so that at length, running through the whole circle of transformations, and completing its period, it in some degree restores itself, if the force be continued. And that method of binding, torturing, or detaining, will prove the most effectual and expeditious, which makes use of manacles and fetters; that is, lays hold and works upon matter in the extremest degrees.

The addition in the fable that makes Proteus a prophet, who had the knowledge of things past, present, and future, excellently agrees with the nature of matter; as he who knows the properties, the changes, and the processes of matter, must of necessity understand the effects and sum of what it does, has done, or can do, though his knowledge extends not to all the parts and particulars thereof.

JOHN DONNE
(1572–1631)

Loves Alchymie

Some that have deeper digg'd loves Myne then I,
Say, where his centrique happinesse doth lie:
   I have lov'd, and got, and told,
But should I love, get, tell, till I were old,
I should not finde that hidden mysterie;
   Oh, 'tis imposture all:
And as no chymique yet th'Elixar got,
   But glorifies his pregnant pot,
   If by the way to him befall
Some odoriferous thing, or medicinall,
   So, lovers dreame a rich and long delight,
   But get a winter-seeming summers night.

Our ease, our thrift, our honour, and our day,
Shall we, for this vaine Bubles shadow pay?
   Ends love in this, that my man,
Can be as happy as I can; If he can
Endure the short scorne of a Bridegroomes play?
   That loving wretch that sweares,
'Tis not the bodies marry, but the mindes,
   Which he in her Angelique findes,
   Would sweare as justly, that he heares,
In that dayes rude hoarse minstralsey, the spheares.
   Hope not for minde in women; at their best
Sweetnesse and wit, they'are but *Mummy* possest.

BEN JONSON

(1572–1637)

To Alchymists

IF all you boast of your great art be true;
Sure, willing povertie lives most in you.

From *The Alchemist*

[In the following scenes Sir Epicure Mammon, an avaricious knight, waits with great impatience for the philosopher's stone, which the alchemist, Subtle, has promised him. Mammon is a man possessed, dreaming "he'll turn the world, to gold." The world, as his lusts betray, is a shabby place filled with rogues and gamblers like Surly, who is skeptical of alchemy, and Face, the alchemist's assistant, who is a master at the delay tactics necessary to trap a gull. The characters in Jonson's play are well-matched tricksters all skilled in artful jargon. It is not necessary to understand all the terms of their colorful language to catch the essence of their encounter—which is a kind of a street game like Three-Card Monte.]

Characters

FACE, *the alchemist's assistant*
MAMMON, *a knight for whom the alchemist is performing a transmutation*

SUBTLE, *the alchemist*
SURLY, *a gambler skeptical of alchemy but also interested in gold*

ACT II, SCENE ii

[*Enter* FACE]

MAMMON	Do we succeed? Is our day come? And holds it?
FACE	The evening will set red, upon you, sir;
	You have colour for it, crimson: the red ferment
	Has done his office. Three hours hence, prepare you
	To see projection.[1]
MAMMON	Pertinax, my Surly,
	Again, I say to thee, aloud: be rich.
	This day, thou shalt have ingots: and tomorrow,
	Give lords th'affront. Is it, my Zephyrus, right?
	Blushes the bolt's head?
FACE	Like a wench with child, sir,
	That were, but now, discovered to her master.
MAMMON	Excellent witty Lungs! My only care is,
	Where to get stuff, enough now, to project on,
	This town will not half serve me.
FACE	No, sir? Buy
	The covering off o'churches.[2]
MAMMON	That's true.
FACE	Yes.
	Let 'em stand bare, as do their auditory.
	Or cap 'em, new, with shingles.
MAMMON	No, good thatch:
	Thatch will lie light upo' the rafters, Lungs.
	Lungs, I will manumit[3] thee, from the furnace;
	I will restore, thee thy complexion, Puff,
	Lost in the embers; and repair this brain,
	Hurt wi' the fume o' the metals.

---

[1] The final stage in alchemy.
[2] Lead roofs.
[3] Release.

FACE   I have blown,
      sir,
Hard, for your worship; thrown by many a coal,
When 'twas not beech; weighted those I put in, just,
To keep your heat, still even; these bleared eyes
Have waked, to read your several colours, sir,
Of the pale citron, the green lion, the crow,
The peacock's tail, the plumed swan.[4]

MAMMON   And, lastly,
Thou hast descried the flower, the *sanguis agni*?

FACE   Yes, sir.

MAMMON   Where's master?

FACE   At's prayers, sir, he,
Good man, he's doing his devotions,
For the success.

MAMMON   Lungs, I will set a period,
To all thy labours: thou shalt be the master
Of my seraglio.

FACE   Good, sir.

MAMMON   But do you hear?
I'll geld you, Lungs.

FACE   Yes, sir.

MAMMON   For I do mean
To have a list of wives, and concubines,
Equal with Solomon; who had the stone
Alike, with me: and I will make me, a back
With the elixir, that shall be as tough
As Hercules, to encounter fifty a night.[5]
Th'art sure, thou saw'st it blood?

FACE   Both blood, and spirit, sir.

MAMMON   I will have all my beds, blown up; not stuffed:
Down is too hard. And then, mine oval room,
Filled with such pictures, as Tiberius took
From Elephantis: and dull Aretine

[4]The colors indicate stages of fermentation ascending toward red, *sanguis agni,* the final color of projection.
[5]King Thespius gave Hercules the gift of fifty nights with fifty daughters in gratitude for his slaying of the lion of Cytheron.

But coldly imitated.[6] Then, my glasses,
Cut in more subtle angles, to disperse,
And multiply the figures, as I walk
Naked between my succubae. My mists
I'll have of perfume, vapoured 'bout the room,
To lose ourselves in; and my baths, like pits
To fall into: from whence, we will come forth,
And roll us dry in gossamer, and roses.
(Is it arrived at ruby?)—Where I spy
A wealthy citizen, or rich lawyer,
Have a sublimed pure wife, unto that fellow
I'll send a thousand pound, to be my cuckold.

FACE   And I shall carry it?

MAMMON   No. I'll ha' no bawds,
But fathers, and mothers. They will do it best.
Best of all others. And, my flatterers
Shall be the pure, and gravest of Divines,
That I can get for money. My mere fools,
Eloquent burgesses, and then my poets
The same that writ so subtly of the fart,
Whom I will entertain, still, for that subject.
The few, that would give out themselves, to be
Court, and town stallions, and, each-where,
    belie
Ladies, who are known most innocent, for
    them;
Those will I beg, to make me eunuchs of:
And they shall fan me with ten ostrich tails
Apiece, made in a plume, to gather wind.
We will be brave, Puff, now we ha' the
    medicine.
My meat, shall all come in, in Indian shells,
Dishes of agate, set in gold, and studded,
With emeralds, sapphires, hyacinths, and
    rubies.
The tongues of carps, dormice, and camels'
    heels,

---

[6] Pietro Aretino (1492–1556) wrote sixteen sonnets, *Sonnetti lussuriosi* (1523), as companions for Giulio Romano's obscene designs.

Boiled i' the spirit of Sol,[7] and dissolved pearl,
(Apicius' diet,[8] 'gainst the epilepsy)
And I will eat these broths, with spoons of
    amber,
Headed with diamond, and carbuncle.
My footboy shall eat pheasants, calvered
    salmons,
Knots, godwits, lampreys: I myself will have
The beards of barbels, served, instead of salads;
Oiled mushrooms; and the swelling unctuous
    paps
Of a fat pregnant sow, newly cut off,
Dressed with an exquisite, and poignant sauce;
For which, I'll say unto my cook, there's gold,
Go forth, and be a knight.

FACE   Sir, I'll go look
A little, how it heightens.

MAMMON   Do. My shirts
I'll have of taffeta-sarsnet, soft, and light
As cobwebs; and for all my other raiment
It shall be such, as might provoke the Persian;
Were he to teach the world riot, anew.
My gloves of fishes', and birds' skins, perfumed
With gums of paradise, and eastern air—

SURLY   And do you think to have the stone, with this?

MAMMON   No, I do think, t'have all this, with the stone.

SURLY   Why, I have heard, he must be *homo frugi*,[9]
A pious, holy, and religious man,
One free from mortal sin, a very virgin.

MAMMON   That makes it, sir, he is so. But I buy it.
My venture brings it me. He, honest wretch,
A notable, superstitious, good soul,
Has worn his knees bare, and his slippers bald,
With prayer, and fasting for it: and, sir, let him

[7] Gold.
[8] A famous Roman gourmand.
[9] Alchemical treatises insisted that apprentices to the art lead pious and austere lives. When Subtle, the alchemist, appears in the next scene, he is addressed in priestly terms, as Father, and religious language is used throughout.

Do it alone, for me, still. Here he comes,
Not a profane word, afore him: 'tis poison.

ACT II, SCENE iii

[*Enter* SUBTLE]

MAMMON   Good morrow, Father.

SUBTLE   Gentle son, good morrow,
And, to your friend, there. What is he, is with
     you?

MAMMON   An heretic, that I did bring along,
In hope, sir, to convert him.

SUBTLE   Son, I doubt
You're covetous, that thus you meet your time
I' the just point: prevent your day, at morning.
This argues something, worthy of a fear
Of importune, and carnal appetite.
Take heed, you do not cause the blessing to
     leave you,
With your ungoverened haste. I should be
     sorry,
To see my labours, now, e'en at perfection,
Got by long watching, and large patience,
Not prosper, where my love, and zeal hath
     placed 'em.
Which (heaven I call to witness, with yourself,
To whom, I have poured my thoughts) in all my
     ends,
Have looked no way, but unto public good,
To pious uses, and dear charity,
Now grown a prodigy with men. Wherein
If you, my son, should now prevaricate,
And, to your own particular lusts, employ
So great, and catholic a bliss: be sure,
A curse will follow, yea, and overtake
Your subtle, and most secret ways.

MAMMON   I know, sir,
You shall not need to fear me. I but come,
To ha' you confute this gentleman.

SURLY   Who is,
Indeed, sir, somewhat costive of belief
Toward your stone:[10] would not be gulled.
SUBTLE   Well, son,
All that I can convince him in, is this,
The work is done: bright Sol is in his robe.
We have a medicine of the triple soul,
The glorified spirit.[11] Thanks be to heaven,
And make us worthy of it. Eulenspiegel.[12]
[*Enter* FACE]
FACE   Anon, sir.
SUBTLE   Look well to the register,[13]
And let your heat, still, lessen by degrees,
To the aludels.[14]
FACE   Yes sir.
SUBTLE   Did you look
O' the bolt's head yet?
FACE   Which, on D,[15] sir?
SUBTLE   Ay.
What's the complexion?
FACE   Whitish.
SUBTLE   Infuse vinegar,
To draw his volatile substance, and his tincture:
And let the water in glass E be filtered,
And put into the gripe's egg.[16] Lute him well;[17]
And leave him closed in *balneo*.[18]
FACE   I will, sir.
SURLY   What a brave language here is? Next to canting?
SUBTLE   I have another work; you never saw, son,
That, three days since, passed the philosopher's
        wheel,
In the lent heat of Athanor; and's become
Sulphur o'nature.

[10] The philosopher's stone.
[11] The elixir.
[12] "Owl glass," a folkloric name for a jesting knave.
[13] Damper.
[14] Special pots with two openings.
[15] Letters refer to different furnaces.
[16] A pot in the shape of a griffin's egg.
[17] "Cover with clay."
[18] "The bath."

MAMMON   But 'tis for me?

SUBTLE   What need you?
You have enough, in that is, perfect.

MAMMON   O, but—

SUBTLE   Why, this is covetise!

MAMMON   No, I assure you,
I shall employ it all, in pious uses,
Founding of colleges, and grammar schools,
Marrying young virgins, building hospitals,
And now, and then, a church.

SUBTLE   How now?

FACE   Sir, please you,
Shall I not change the filter?

SUBTLE   Marry, yes.
And bring me the complexion of glass B.

[*Exit* FACE]

MAMMON   Ha' you another?

SUBTLE   Yes, son, were I assured
Your piety were firm, we would not want
The means to glorify it. But I hope the best:
I mean to tinct C in sand-heat, tomorrow,
And give him imbibition.[19]

MAMMON   Of white oil?

SUBTLE   No, sir, of red. F is come over to the helm too,
I thank my Maker, in S. Mary's bath,[20]
And shows *lac virginis.*[21] Blessed be heaven
I sent you of his faeces there, calcined.
Out of the calx,[22] I ha' won the salt of mercury.

MAMMON   By pouring on your rectified water?

SUBTLE   Yes, and reverberating in Athanor.

[*Enter* FACE]

How now? What colour says it?

FACE   The ground black, sir.

MAMMON   That's your crow's head?[23]

SURLY   Your cockscomb's, is't not?

[19]"Soaking."
[20]"Heat."
[21]"Water of mercury."
[22]"Burnt powder."
[23]A symbol of total calcination.

SUBTLE   No, 'tis not perfect, would it were the crow.
             That work wants something.

SURLY   (O, I looked for this.
             The hay is a-pitching.)

SUBTLE   Are you sure, you loosed 'em
             I' their own menstrue?

FACE   Yes, sir, and then married 'em,
             And put 'em in a bolt's head, nipped to
                 digestion,
             According as you bade me; when I set
             The liquor of Mars[24] to circulation,
             In the same heat.

SUBTLE   The process, then, was right.

FACE   Yes, by the token, sir, the retort broke,
             And what was saved, was put into the
                 pelican,[25]
             And signed with Hermes' seal.[26]

SUBTLE   I think 'twas so.
             We should have a new amalgama.[27]

SURLY   O, this ferret
             Is rank as any pole-cat.

SUBTLE   But I care not.
             Let him e'en die; we have enough beside,
             In embrion[28] H has his white shirt on?

FACE   Yes, sir,
             He's ripe for inceration: he stands warm,
             In his ash-fire. I would not, you should let
             Any die now, if I might counsel, sir,
             For luck's sake to the rest It is not good.

MAMMON   He says right.

SURLY   Ay, are you bolted?

FACE   Nay, I knows't, sir,
             I have seen th' ill fortune. What is some three
                 ounces
             Of fresh materials?

---

[24] Molten iron wisely named for the god of war.
[25] A vessel shaped like the bird.
[26] "Hermetically sealed."
[27] A mixture of mercury with unspecified metals.
[28] Metal in combination.

MAMMON   Is't no more?

FACE   No more, sir,
Of gold, t'amalgam, with some six of mercury.

MAMMON   Away, here's money. What will serve?

FACE   Ask him, sir.

MAMMON   How much?

SUBTLE   Give him nine pound: you may gi' him ten.

SURLY   Yes, twenty, and be cozened, do.

MAMMON   There 'tis.

SUBTLE   This needs not. But that you will have it, so,
To see conclusions of all. For two
Of our inferior works, are at fixation.[29]
A third is in ascension.[30] Go your ways.
Ha' you set the oil of Luna in kemia?[31]

FACE   Yes, sir.

SUBTLE   And the philosopher's vinegar?

FACE   Ay.

[*Exit* FACE]

SURLY   We shall have a salad.

MAMMON   When do you make projection?

SUBTLE   Son, be not hasty, I exalt our medicine,
By hanging him in *balneo vaporoso;*
And giving him solution; then congeal him;
And then dissolve him; then again congeal him;
For look, how oft I iterate the work,
So many times, I add unto his virtue.
As, if at first, one ounce convert a hundred,
After his second loose, he'll turn a thousand;
His third solution, ten; his fourth, a hundred.
After his fifth, a thousand thousand ounces
Of any imperfect metal, into pure
Silver, or gold, in all examinations,
As good, as any of the natural mine.
Get you your stuff here, against afternoon,
Your brass, your pewter, and your andirons.

MAMMON   Not those of iron?

[29] Reduced from volatile to solid form.
[30] Distillation.
[31] "Put the white elixir into the gourd-shaped retort."

SUBTLE	Yes, you may bring them, too. We'll change all metals.
SURLY	I believe you, in that.
MAMMON	Then I may send my spits?
SUBTLE	Yes, and your racks.
SURLY	And dripping pans, and pot-hangers, and hooks? Shall he not?
SUBTLE	If he please.
SURLY	To be an ass.
SUBTLE	How, sir!
MAMMON	This gentleman, you must bear withal. I told you, he had no faith.
SURLY	And little hope, sir, But, much less charity, should I gull myself.
SUBTLE	Why, what have you observed, sir, in our art, Seems so impossible?
SURLY	But your whole work, no more. That you should hatch gold in a furnace, sir, As they do eggs, in Egypt![32]
SUBTLE	Sir, do you Believe that eggs are hatched so?
SURLY	If I should?
SUBTLE	Why, I think that the greater miracle. No egg, but differs from a chicken, more, Than metals in themselves.
SURLY	That cannot be. The egg's ordained by nature, to that end: And is a chicken in *potentia*.
SUBTLE	The same we say of lead, and other metals, Which would be gold, if they had time.
MAMMON	And that Our art doth further.
SUBTLE	Ay, for 'twere absurd To think that nature, in the earth, bred gold Perfect, i' the instant. Something went before. There must be remote matter.

[32] Pliny describes Egyptian incubation ovens. That metal strives to become gold as eggs strive to become chickens is classical alchemical theory.

SURLY   Ay, what is that?
SUBTLE   Marry, we say—
MAMMON   Ay, now it heats: stand Father.
            Pound him to dust—
SUBTLE   It is, of the one part,
            A humid exhalation, which we call
            Materia liquida, or the unctuous water;
            On th' other part, a certain crass, and viscous
            Portion of earth; both which, concorporate,
            Do make the elementary matter of gold:
            Which is not, yet, *propria materia,*
            But common to all metals, and all stones.
            For, where it is forsaken of that moisture
            And hath more dryness, it becomes a stone;
            Where it retains more of the humid fatness,
            It turns to sulphur, or to quicksilver:
            Who are the parents of all other metals.
            Nor can this remote matter, suddenly,
            Progress so from extreme, unto extreme,
            As to grow gold, and leap o'er all the means.
            Nature doth, first, beget th' imperfect; then
            Proceeds she to the perfect. Of that airy,
            And oily water, mercury is engendered;
            Sulphur o' the fat, and earthy part: the one
            (Which is the last) supplying the place of the
                    male,
            The other of the female, in all metals.
            Some do believe hermaphrodeity,
            That both do act, and suffer. But, these two
            Make the rest ductile, malleable, extensive.
            And, even in gold, they are; for we do find
            Seeds of them, by our fire, and gold in them:
            And can produce the species of each metal
            More perfect thence, than nature doth in earth.
            Beside, who doth not see, in daily practice,
            Art can beget bees, hornets, beetles, wasps,
            Out of the carcasses, and dung of creatures;
            Yea, scorpions, of an herb, being rightly placed:
            And these are living creatures, far more perfect,
            And excellent, than metals.

MAMMON   Well said, Father!
    Nay, if he take you in hand, sir, with an
        argument,
    He'll bray you in a mortar

SURLY   Pray you, sir, stay.
    Rather, than I'll be brayed, sir, I'll believe,
    That alchemy is a pretty kind of game,
    Somewhat like tricks o' the cards, to cheat a
        man,
    With charming.

SUBTLE   Sir?

SURLY   What else are all your terms,
    Whereon no one o' your writers 'grees with
        other?
    Of your elixir, your *lac virginis,*
    Your stone, your medicine, and your
        chrysosperm,[33]
    Your sal, your sulphur, and your mercury,
    Your oil of height, your tree of life, your blood,
    Your marcasite, your tutty,[34] your magnesia.
    Your toad, your crow, your dragon, and your
        panther,
    Your sun, your moon, our firmament, your
        adrop,[35]
    Your lato, azoch, zernich, chibrit, autarit,[36]
    And then, your red man, and your white
        woman,[37]
    With all your broths, your menstrues, and
        materials,
    Of piss, and eggshells, women's terms, man's
        blood,
    Hair o' the head, burnt clouts, chalk, merds,
        and clay,
    Powder of bones, scalings of iron, glass,
    And worlds of other strange ingredients,
    Would burst a man to name?

[33] "Seed of gold."
[34] "Impure zinc"
[35] "Lead."
[36] Various metals and compounds.
[37] "Sulphur and mercury."

SUBTLE   And all these, named
Intending but one thing: which art our writers
Used to obscure their art.

MAMMON   Sir, so I told him,
Because the simple idiot should not learn it,
And make it vulgar.

SUBTLE   Was not all the knowledge
Of the Egyptians writ in mystic symbols?
Speak not the Scriptures, oft, in parables?
Are not the choicest fables of the poets,
That were the fountains, and first springs of
     wisdom,
Wrapped in perplexed allegories?

PERCY BYSSHE SHELLEY

(1792–1822)

From "Alastor: or, The Spirit of Solitude"

O, for Medea's wondrous alchymy,
Which whereso'er it fell made the earth gleam
With bright flowers, and the wintry boughs exhale
From vernal blooms fresh fragrance! O, that God,
Profuse of poison, would concede the chalice
Which but one living man has drained, who now,
Vessel of deathless wrath, a slave that feels
No proud exemption in the blighting curse
He bears, over the world wanders for ever,
Lone as incarnate death! O, that the dream
Of dark magician in his visioned cave,

Raking the cinders of a crucible
For life and power, even when his feeble hand
Shakes in its last decay, were the true law
Of this so lovely world! But thou art fled
Like some frail exhalation, which the dawn
Robes on its golden beams,—ah, thou hast fled!

MARY SHELLEY
(1797–1851)

From *Frankenstein*, Chapter 3

The next morning I delivered my letters of introduction and paid a visit to some of the principal professors. Chance—or rather the evil influence, the Angel of Destruction, which asserted omnipotent sway over me from the moment I turned my reluctant steps from my father's door—led me first to M. Krempe, professor of natural philosophy. He was an uncouth man, but deeply imbued in the secrets of his science. He asked me several questions concerning my progress in the different branches of science appertaining to natural philosophy. I replied carelessly, and partly in contempt, mentioned the names of my alchemists as the principal authors I had studied. The professor stared. "Have you," he said, "really spent your time in studying such nonsense?"

I replied in the affirmative. "Every minute," continued M. Krempe with warmth, "every instant that you have wasted on those books is utterly and entirely lost. You have burdened your memory with exploded systems and useless names. Good God! In what desert land have you lived, where no one was kind enough to inform you that these fancies which you have so greedily imbibed are a thousand years old and as

musty as they are ancient? I little expected, in this enlightened
and scientific age, to find a disciple of Albertus Magnus and
Paracelsus. My dear sir, you must begin your studies entirely
anew."

So saying, he stepped aside and wrote down a list of several books treating of natural philosophy which he desired me
to procure, and dismissed me after mentioning that in the beginning of the following week he intended to commence a
course of lectures upon natural philosophy in its general relations, and that M. Waldman, a fellow professor, would lecture
upon chemistry the alternate days that he omitted.

I returned home not disappointed, for I have said that I
had long considered those authors useless whom the professor reprobated; but I returned not at all the more inclined
to recur to these studies in any shape. M. Krempe was a little
squat man with a gruff voice and a repulsive countenance; the
teacher, therefore, did not prepossess me in favour of his pursuits. In rather a too philosophical and connected a strain,
perhaps, I have given an account of the conclusions I had
come to concerning them in my early years. As a child I had
not been content with the results promised by the modern
professors of natural science. With a confusion of ideas only
to be accounted for by my extreme youth and my want of a
guide on such matters, I had retrod the steps of knowledge
along the paths of time and exchanged the discoveries of recent inquirers for the dreams of forgotten alchemists. Besides,
I had a contempt for the uses of modern natural philosophy.
It was very different when the masters of the science sought
immortality and power; such views, although futile, were
grand; but now the scene was changed. The ambition of the
inquirer seemed to limit itself to the annihilation of those visions on which my interest in science was chiefly founded. I
was required to exchange chimeras of boundless grandeur for
realities of little worth.

Such were my reflections during the first two or three days
of my residence at Ingolstadt, which were chiefly spent in becoming acquainted with the localities and the principal residents in my new abode. But as the ensuing week commenced,
I thought of the information which M. Krempe had given me

concerning the lectures. And although I could not consent to go and hear that little conceited fellow deliver sentences out of a pulpit, I recollected what he had said of M. Waldman, whom I had never seen, as he had hitherto been out of town.

Partly from curiosity and partly from idleness, I went into the lecturing room, which M. Waldman entered shortly after. This professor was very unlike his colleague. He appeared about fifty years of age, but with an aspect expressive of the greatest benevolence; a few grey hairs covered his temples, but those at the back of his head were nearly black. His person was short but remarkably erect and his voice the sweetest I had ever heard. He began his lecture by a recapitulation of the history of chemistry and the various improvements made by different men of learning, pronouncing with fervour the names of the most distinguished discoverers. He then took a cursory view of the present state of the science and explained many of its elementary terms. After having made a few preparatory experiments, he concluded with a panegyric upon modern chemistry, the terms of which I shall never forget:

"The ancient teachers of this science," said he, "promised impossibilities and performed nothing. The modern masters promise very little; they know that metals cannot be transmuted and that the elixir of life is a chimera. But these philosophers, whose hands seem only made to dabble in dirt, and their eyes to pore over the microscope or crucible, have indeed performed miracles. They penetrate into the recesses of nature and show how she works in her hiding-places. They ascend into the heavens; they have discovered how the blood circulates, and the nature of the air we breathe. They have acquired new and almost unlimited powers; they can command the thunders of heaven, mimic the earthquake, and even mock the invisible world with its own shadows."

Such were the professor's words—rather let me say such the words of the fate—enounced to destroy me. As he went on I felt as if my soul were grappling with a palpable enemy; one by one the various keys were touched which formed the mechanism of my being; chord after chord was sounded, and soon my mind was filled with one thought, one conception, one purpose. So much has been done, exclaimed the soul of

Frankenstein—more, far more, will I achieve; treading in the steps already marked, I will pioneer a new way, explore unknown powers, and unfold to the world the deepest mysteries of creation.

I closed not my eyes that night. My internal being was in a state of insurrection and turmoil; I felt that order would thence arise, but I had no power to produce it. By degrees, after the morning's dawn, sleep came. I awoke, and my yesternight's thoughts were as a dream. There only remained a resolution to return to my ancient studies and to devote myself to a science for which I believed myself to possess a natural talent. On the same day I paid M. Waldman a visit. His manners in private were even more mild and attractive than in public, for there was a certain dignity in his mien during his lecture which in his own house was replaced by the greatest affability and kindness. I gave him pretty nearly the same account of my former pursuits as I had given to his fellow professor. He heard with attention the little narration concerning my studies and smiled at the names of Cornelius Agrippa and Paracelsus, but without the contempt that M. Krempe had exhibited. He said that "These were men to whose indefatigable zeal modern philosophers were indebted for most of the foundations of their knowledge. They had left to us, as an easier task, to give new names and arrange in connected classifications the facts which they in a great degree had been the instruments of bringing to light. The labours of men of genius, however erroneously directed, scarcely ever fail in ultimately turning to the solid advantage of mankind." I listened to his statement, which was delivered without any presumption or affectation, and then added that his lecture had removed my prejudices against modern chemists; I expressed myself in measured terms, with the modesty and deference due from a youth to his instructor, without letting escape (inexperience in life would have made me ashamed) any of the enthusiasm which stimulated my intended labours. I requested his advice concerning the books I ought to procure.

"I am happy," said M. Waldman, "to have gained a disciple; and if your application equals your ability, I have no doubt of your success. Chemistry is that branch of natural philosophy in which the greatest improvements have been and may be made; it is on that account that I have made it my peculiar study; but at the same time, I have not neglected the other branches of science. A man would make but a very sorry chemist if he attended to that department of human knowledge alone. If your wish is to become really a man of science and not merely a petty experimentalist, I should advise you to apply to every branch of natural philosophy, including mathematics."

He then took me into his laboratory and explained to me the uses of his various machines, instructing me as to what I ought to procure and promising me the use of his own when I should have advanced far enough in the science not to derange their mechanism. He also gave me the list of books which I had requested, and I took my leave.

Thus ended a day memorable to me; it decided my future destiny.

ERASMUS DARWIN
(1731–1802)

From *Botanic Garden*

PART I
THE ECONOMY OF VEGETATION

Canto II

Hence dusky *Iron* sleeps in dark abodes,
And ferny foliage nestles in the nodes;
Till with wide lungs the panting bellows blow,
And waked by fire the glittering torrents flow;
—Quick whirls the wheel, the ponderous hammer falls,

Loud anvils ring amid the trembling walls,
Strokes follow strokes, the sparkling ingot shines,
Flows the red slag, the lengthening bar refines;
Cold waves, immersed, the glowing mass congeal,
And turn to adamant the hissing Steel.

Last Michel's hands,[38] with touch of potent charm,
The polish'd rods with powers magnetic arm;
With points directed to the polar stars,
In one long line extend the temper'd bars;
Then thrice and thrice with steady eye he guides,
And o'er the adhesive train the magnet slides;
The obedient Steel with living instinct moves,
And veers for ever to the pole it loves.

Hail, adamantine *Steel!* magnetic Lord!
King of the prow, the plowshare, and the sword!
True to the pole, by thee the pilot guides
His steady helm amid the struggling tides,
Braves with broad sail the immeasurable sea,
Cleaves the dark air, and asks no star but Thee.—
By thee the plowshare rends the matted plain,
Inhumes in level rows the living grain;
Intrusive forests quit the cultured ground,
And Ceres laughs with golden fillets crown'd.—
O'er restless realms when scowling Discord flings
Her snakes, and loud the din of battle rings;
Expiring Strength, and vanquish'd Courage feel
Thy arm resistless, adamantine *Steel*.

Hence in fine streams diffusive *Acids* flow,
Or wing'd with fire o'er Earth's fair bosom blow;
Transmute to glittering Flints her chalky lands,
Or sink on Ocean's bed in countless Sands.
Hence silvery Selenite her crystal moulds,
And soft Asbestus smooths his silky folds;

[38]Reverend Michel, who, in 1750, published a new method of magnetizing steel bars. This method intrigued Darwin because of its suggestions about the relationship between magnetic polarity and the earth's rotation.

His cubic forms phosphoric Fluor prints,
Or rays in spheres his amethystine tints.
Soft cobweb clouds transparent Onyx spreads,
And playful Agates weave their colour'd threads;
Gay pictured Mochoes glow with landscape-dyes,
And changeful Opals roll their lucid eyes;
Blue lambent light around the Sapphire plays,
Bright Rubies blush, and living Diamonds blaze.

*   *   *

Hence ductile *Clays* in wide expansion spread,
Soft as the Cygnet's down, their snow-white bed;
With yielding flakes successive forms reveal,
And change obedient to the whirling wheel.
First China's sons, with early art elate,
Form'd the gay tea-pot, and the pictured plate;
Saw with illumined brow and dazzled eyes
In the red stove vitrescent colours rise;
Speck'd her tall beakers with enamel'd stars,
Her monster-josses,[39] and gigantic jars;
Smear'd her huge dragons with metallic hues,
With golden purples, and cobaltic blues;
Bade on wide hills her porcelain castles glare,
And glazed Pagodas tremble in the air.

*   *   *

Canto I

Nymphs! you disjoin, unite, condense, expand,
And give new wonders to the Chemist's hand;
On tepid clouds of rising steam aspire,
Or fix in sulphur all its solid fire;[40]
With boundless spring elastic airs unfold,
Or fill the fine vacuities of gold;
With sudden flash vitrescent sparks reveal,
By fierce collision from the flint and steel;

*   *   *

[39]Chinese idols.
[40]A reference to current theories of chemical explosion.

You led your FRANKLIN to your glazed retreats,
Your air-built castles, and your silken seats;
Bade his bold arm invade the lowering sky,
And seize the tip-toe lightnings ere they fly;
O'er the young Sage your mystic mantle spread,
And wreathed the crown electric round his head.—
Thus, when on wanton wing intrepid *Love*
Snatch'd the raised lightning from the arm of JOVE:
Quick o'er his knee the triple bolt He bent,
The cluster'd darts and forky arrows rent,
Snapt with illumin'd hands each flaming shaft,
His tingling fingers shook, and stamp'd, and laugh'd;
Bright o'er the floor the scatter'd fragments blazed,
And gods, retreating, trembled as they gazed;
The immortal Sire, indulgent to his child,
Bow'd his ambrosial locks, and Heaven, relenting, smiled.

BENJAMIN FRANKLIN
(1706–90)

From *Works of Benjamin Franklin*

So wonderfully are these two states of electricity, the *plus* and *minus*, combined and balanced in this miraculous bottle! situated and related to each other in a manner that I can by no means comprehend! If it were possible that a bottle should in one part contain a quantity of air strongly compressed, and in another part a perfect vacuum, we know the equilibrium would be instantly restored *within*. But here we have a bottle containing at the same time a *plenum* of electrical fire, and a *vacuum* of the same fire; and yet the equilibrium cannot be restored between them but by a communication *without!*

though the *plenum* presses violently to expand, and the hungry vacuum seems to attract as violently in order to be filled.

EXPERIMENT VI

Place a man on a cake of wax, and present him the wire of the electrified phial to touch, you standing on the floor, and holding it in your hand. As often as he touches it, he will be electrified *plus;* and any one standing on the floor may draw a spark from him. The fire in this experiment passes out of the wire into him; and at the same time out of your hand into the bottom of the bottle.

EXPERIMENT VII

Give him the electrical phial to hold; and do you touch the wire; as often as you touch it, he will be electrified *minus,* and may draw a spark from any one standing on the floor. The fire now passes from the wire to you, and from him into the bottom of the bottle.

*   *   *

EXPERIMENT X

Though, as in *Experiment* VI, a man standing on wax may be electrized a number of times by repeatedly touching the wire of an electrized bottle (held in the hand of one standing on the floor), he receiving the fire from the wire each time; yet holding it in his own hand, and touching the wire, though he draws a strong spark, and is violently shocked, no electricity remains in him; the fire only passing through him, from the upper to the lower part of the bottle. Observe, before the shock, to let some one on the floor touch him to restore the equilibrium in his body; for, in taking hold of the bottom of the bottle, he sometimes becomes a little electrized *minus,* which will continue after the shock, as would any *plus* electricity, which he might have given him before the shock. For restoring the equilibrium in the bottle does not at all affect the electricity in the man through whom the fire passes; that electricity is neither increased nor diminished.

*   *   *

Chagrined a little that we have been hitherto able to produce nothing in this way of use to mankind; and the hot weather coming on, when electrical experiments are not so agreeable, it is proposed to put an end to them for this season, somewhat humorously, in a party of pleasure on the banks of the *Skuylkill.*[41] Spirits, at the same time, are to be fired by a spark sent from side to side through the river, without any other conductor than the water; an experiment which we some time since performed, to the amazement of many. A turkey is to be killed for our dinner by the *electrical shock,* and roasted by the *electrical jack,* before a fire kindled by the *electrified bottle;* when the healths of all the famous electricians in England, Holland, France, and Germany are to be drank in *electrified bumpers,* under the discharge of guns from the *electrical battery.*

TO CADWALLDER COLDEN, AT NEW YORK.
COMMUNICATED TO MR. COLLINSON.
Unlimited Nature of the Electric Force.

Philadelphia, 1751

. . . I forget whether I wrote to you, that I have melted brass pins and steel needles, inverted the poles of the magnetic needle, given a magnetism and polarity to needles that had none, and fired dry gunpowder by the electric spark. I have five bottles that contain eight or nine gallons each, two of which charged are sufficient for those purposes; but I can charge and discharge them altogether. There are no bounds (but what expense and labor give) to the force man may raise and use in the electrical way; for bottle may be added to bottle *in infinitum,* and all united and discharged together as one, the force and effect proportioned to their number and size. The greatest known effects of common lightning may, I think, without much difficulty, be exceeded in this way, which a few years since

[41]The Philadelphia river.

could not have been believed, and even may now seem to many a little extravagant to suppose. So we are got beyond the skill of Rabelais's devils of two years old, who, he humorously says, had only learned to thunder and lighten a little round the head of a cabbage.

I am, with sincere respect,
Your most obliged humble servant,
B. Franklin.

EMILY DICKINSON
(1830–86)

The Farthest Thunder That I Heard

The farthest Thunder that I heard
Was nearer than the Sky
And rumbles still, though torrid Noons
Have lain their missiles by—
The Lightning that preceded it
Struck no one but myself—
But I would not exchange the Bolt
For all the rest of Life—
Indebtedness to Oxygen
The Happy may repay,
But not the obligation
To Electricity—
It founds the Homes and decks the Days
And every clamor bright
Is but the gleam concomitant
Of that waylaying Light—
The Thought is quiet as a Flake—
A Crash without a Sound,
How Life's reverberation
It's Explanation found—

SIR HUMPHRY DAVY
(1778–1829)

From *Researches, Chemical and Philosophical*

After I had taken a situation in which I could by means of a curved thermometer inserted under the arm, and a stop-watch, ascertain the alterations in my pulse and animal heat, 20 quarts of nitrous oxide were thrown into the box.

For three minutes I experienced no alteration in my sensations, though immediately after the introduction of the nitrous oxide the smell and taste of it were very evident.

In four minutes I began to feel a slight glow in the cheeks, and a generally diffused warmth over the chest, though the temperature of the box was not quite 50°. I had neglected to feel my pulse before I went in; at this time it was 104 and hard, the animal heat was 98°. In ten minutes the animal heat was near 99°, in a quarter of an hour 99.5°, when the pulse was 102, and fuller than before.

At this period 20 quarts more of nitrous oxide were thrown into the box, and well mingled with the mass of air by agitation.

In 25 minutes the animal heat was 100°, pulse 124. In 30 minutes, 20 quarts more of gas were introduced.

My sensations were now pleasant; I had a generally diffused warmth without the slightest moisture of the skin, a sense of exhilaration similar to that produced by a small dose of wine, and a disposition to muscular motion and to merriment.

In three quarters of an hour the pulse was 104, and animal heat not quite 99.5°, the temperature of the chamber was 64°. The pleasurable feelings continued to increase, the pulse became fuller and slower, till in about an hour it was 88, when the animal heat was 99°.

20 quarts more of air were admitted. I had now a great disposition to laugh; luminous points seemed frequently to pass before my eyes, my hearing was certainly more acute, and I felt a pleasant lightness and power of exertion in my muscles. In a short time the symptoms became stationary; breathing was rather oppressed, and on account of the great desire of action, rest was painful.

I now came out of the box, having been in precisely an hour and quarter.

The moment after, I began to respire 20 quarts of unmingled nitrous oxide. A thrilling, extending from the chest to the extremities, was almost immediately produced. I felt a sense of tangible extension highly pleasurable in every limb; my visible impressions were dazzling, and apparently magnified, I heard distinctly every sound in the room, and was perfectly aware of my situation. By degrees, as the pleasurable sensations increased, I lost all connection with external things; trains of vivid visible images rapidly passed through my mind, and were connected with words in such a manner, as to produce perceptions perfectly novel. I existed in a world of newly connected and newly modified ideas. I theorised—I imagined that I made discoveries. When I was awakened from this semidelirious trance by Dr. Kinglake, who took the bag from my mouth, indignation and pride were the first feelings produced by the sight of the persons about me. My emotions were enthusiastic and sublime; and for a minute I walked round the room, perfectly regardless of what was said to me. As I recovered my former state of mind, I felt an inclination to communicate the discoveries I had made during the experiment. I endeavoured to recall the ideas, they were feeble and indistinct; one collection of terms, however, presented itself: and with the most intense belief and prophetic manner, I exclaimed to Dr. Kinglake, "*Nothing exists but thoughts!—the universe is composed of impressions, ideas, pleasures and pains!*"

*　　*　　*

I have often felt very great pleasure when breathing it alone, in darkness and silence, occupied only by ideal existence. In two or three instances when I have breathed it amidst noise, the sense of hearing has been painfully affected even by moderate intensity of sound. The light of the sun has sometimes been disagreeably dazzling. I have once or twice felt an uneasy sense of tension in the cheeks and transient pains in the teeth.

Whenever I have breathed the gas after excitement from moral or physical causes, the delight has been often intense and sublime.

On May 5th, at night, after walking for an hour amidst the scenery of the Avon, at this period rendered exquisitely beautiful by bright moonshine; my mind being in a state of agreeable feeling, I respired six quarts of newly prepared nitrous oxide.

The thrilling was very rapidly produced. The objects around me were perfectly distinct, and the light of the candle not as usual dazzling. The pleasurable sensation was at first local, and perceived in the lips and about the cheeks. It gradually, however, diffused itself over the whole body, and in the middle of the experiment was for a moment so intense and pure as to absorb existence. At this moment, and not before, I lost consciousness; it was, however, quickly restored, and I endeavoured to make a by-stander acquainted with the pleasure I experienced by laughing and stamping. I had no vivid ideas. The thrilling and the pleasurable feeling continued for many minutes; I felt two hours afterwards, a slight recurrence of them, in the intermediate state between sleeping and waking; and I had during the whole of the night, vivid and agreeable dreams. I awoke in the morning with the feeling of restless energy, or that desire of action connected with no definite object, which I had often experienced in the course of experiments in 1799.

*　　*　　*

From the nature of the language of feeling, the preceding detail contains many imperfections; I have endeavoured to give as accurate an account as possible of the strange effects of nitrous oxide, by making use of terms standing for the most similar common feelings.

We are incapable of recollecting pleasures and pains of sense. It is impossible to reason concerning them, except by means of terms which have been associated with them at the moment of their existence, and which are afterwards called up amidst trains of concomitant ideas.

When pleasures or pains are new or connected with new ideas, they can never be intelligibly detailed unless associated during their existence with terms standing for analogous feelings.

I have sometimes experienced from nitrous oxide, sensations similar to no others, and they have consequently been indescribable. This has been likewise often the case with other persons. Of two paralytic patients who were asked what they felt after breathing nitrous oxide, the first answered, *"I do not know how, but very queer."* The second said, *"I felt like the sound of a harp."* Probably in the one case, no analogous feelings had ever occurred. In the other, the pleasurable thrillings were similar to the sensations produced by music; and hence, they were connected with terms formerly applied to music.

To Mr. Davy.
London, Sept. 21st, 1799

Detail of Mr. Coleridge

The first time I inspired the nitrous oxide, I felt a highly pleasurable sensation of warmth over my whole frame, resembling that which I remember once to have experienced after returning from a walk in the snow into a warm room. The only motion which I felt inclined to make, was that of laughing at those who were looking at me. My eyes felt distended, and towards the last, my heart beat as if it were leaping up and down. On removing the mouth-piece, the whole sensation went off almost instantly.

The second time I felt the same pleasurable sensation of warmth, but not, I think, in quite so great a degree. I wished to know what effect it would have on my impressions; I fixed my eye on some trees in the distance, but I did not find any other effect except that they became dimmer and dimmer, and looked at last as if I had seen them through tears. My heart beat more violently than the first time. This was after a hearty dinner.

The third time I was more violently acted on than in the two former. Towards the last, I could not avoid, nor indeed felt any wish to avoid, beating the ground with my feet; and after the mouth-piece was removed, I remained for a few seconds motionless, in great extacy.

The fourth time was immediately after breakfast. The few first impressions affected me so little, that I thought Mr. Davy had given me atmospheric air: but soon felt the warmth beginning about my chest, and spreading upward and downward, so that I could feel its progress over my whole frame. My heart did not beat so violently; my sensations were highly pleasurable, not so intense or apparently local, but of more unmingled pleasure than I had ever before experienced.

S. T. Coleridge

EMILY DICKINSON
(1830–86)

The Chemical Conviction

The Chemical conviction
That Nought be lost
Enable in Disaster
My fractured Trust—

The Faces of the Atoms
If I shall see
How more the Finished Creatures
Departed Me!

JOHN DALTON
(1766–1844)

From *A New System of Chemical Philosophy*

ON THE CONSTITUTION OF BODIES

There are three distinctions in the kinds of bodies, or three states, which have more especially claimed the attention of philosophical chemists; namely, those which are marked by the terms *elastic fluids, liquids,* and *solids.* A very familiar instance is exhibited to us in water, of a body, which, in certain circumstances, is capable of assuming all the three states. In steam we recognise a prefectly elastic fluid, in water, a perfect liquid, and in ice a complete solid. These observations have tacitly led to the conclusion which seems universally adopted, that all bodies of sensible magnitude, whether liquid or solid, are constituted of a vast number of extremely small particles, or atoms of matter bound together by a force of attraction, which is more or less powerful according to circumstances, and which as it endeavors to prevent their separation, is very properly called in that view, *attraction of cohesion;* but as it collects them from a dispersed state (as from steam into water) it is called, *attraction of aggregation,* or more simply, *affinity.* Whatever names it may go by, they still signify one and the same power. It is not my design to call in question this conclusion, which appears completely satisfactory; but to show that we have hitherto made no use of it, and that the consequence of the neglect, has been a very obscure view of chemical agency, which is daily growing more so in proportion to the new lights attempted to be thrown upon it.

The opinions I more particularly allude to, are those of Berthollet[42] on the Laws of chemical affinity; such as that chemical agency is proportional to the mass, and that in all chemical unions, there exist insensible gradations in the proportions of the constituent principles. The inconsistence of these opinions, both with reason and observation, cannot, I think, fail to strike everyone who takes a proper view of the phenomena.

Whether the ultimate particles of a body, such as water, are all alike, that is, of the same figure, weight, etc. is a question of some importance. From what is known, we have no reason to apprehend a diversity in these particulars: if it does exist in water, it must equally exist in the elements con-

---

[42]Claude Louis Berthollet (1748–1800) repeated experiments on elastic fluids done by Lavoisier, Priestly, and Karl Wilhelm Scheele (1742–86). He was a synthesizer of theory who, like Frankenstein, wanted to admit the ideas of Paracelsus to modern chemistry. Later, he became preoccupied with practical applications, experimented on gunpowder, chlorine, and prussic acid, and provided the earliest analysis of ammonia (1785).

stituting water, namely, hydrogen and oxygen. Now it is scarcely possible to conceive how the aggregates of dissimilar particles should be so uniformly the same. If some of the particles of water were heavier than others, if a parcel of the liquid on any occasion were constituted principally of these heavier particles, it must be supposed to affect the specific gravity of the mass, a circumstance not known. Similar observations may be made on other substances. Therefore we may conclude that *the ultimate particles of all homogeneous bodies are perfectly alike in weight, figure, etc.* In other words, every particle of water is like every other particle of water: every particle of hydrogen is like every other particle of hydrogen, etc.

MATTHEW ARNOLD

(1822–88)

From "Empedocles on Etna"

To the elements it came from
Everything will return—
Our bodies to earth,
Our blood to water,
Heat to fire,
Breath to air.
They were well born, they will be well entomb'd—
But mind? . . .

And we might gladly share the fruitful stir
Down in our mother earth's miraculous womb;
Well would it be
With what roll'd of us in the stormy main;
We might have joy, blent with the all-bathing air,
Or with the nimble, radiant life of fire.

But mind, but thought—
If these have been the master part of us—
Where will *they* find their parent element?
What will receive *them*, who will call *them* home?
But we shall still be in them, and they in us,
And we shall be the strangers of the world,
And they will be our lords, as they are now;
And keep us prisoners of our consciousness,
And never let us clasp and feel the All
But through their forms, and modes, and stifling veils.
And we shall be unsatisfied as now;
And we shall feel the agony of thirst,
The ineffable longing for the life of life
Baffled for ever; and still thought and mind
Will hurry us with them on their homeless march,
Over the unallied unopening earth,
Over the unrecognizing sea; while air
Will blow us fiercely back to sea and earth,
And fire repel us from its living waves.
And then we shall unwillingly return
Back to this meadow of calamity,
This uncongenial place, this human life;
And in our individual human state
Go through the sad probation all again,
To see if we will poise our life at last,
To see if we will now at last be true
To our own only true, deep-buried selves,
Being one with which we are one with the whole world;
Or whether we will once more fall away
Into some bondage of the flesh or mind,
Some slough of sense, or some fantastic maze
Forged by the imperious lonely thinking-power.
And watch succeeding age in which we are born
Will have more peril for us than the last;
Will goad our senses with a sharper spur,
Will fret our minds to an intenser play,
Will make ourselves harder to be discern'd.
And we shall struggle awhile, gasp and rebel—
And we shall fly for refuge to past times,

Their soul of unworn youth, their breath of greatness;
And the reality will pluck us back,
Knead us in its hot hand, and change our nature
And we shall feel our powers of effort flag,
And rally them for one last fight—and fail;
And we shall sink in the impossible strife,
And be astray forever.

                         Slave of sense
I have in no wise been;—but slave of thought? . . .
And who can say: I have been always free,
Lived ever in the light of my own soul?—
I cannot; I have liv'd in wrath and gloom,
Fierce, disputatious, ever at war with man,
Far from my own soul, far from warmth and light.
But I have not grown easy in these bonds—
But I have not denied what bonds these were.
Yea, I take myself to witness,
That I have loved no darkness,
Sophisticated no truth,
Nursed no delusion,
Allowed no fear!

And therefore, O ye elements! I know—
Ye know it too—it hath been granted me
Not to die wholly, not to be all enslaved.
I feel it in this hour. The numbing cloud
Mounts off my soul: I feel it, I breathe free.

Is it but for a moment?
—Ah boil up, ye vapours!
Leap and roar, thou sea of fire!
My soul glows to meet you.
Ere it flag, ere the mists
Of despondency and gloom
Rush over it again,
Receive me, save me!
[He plunges into the crater.]
(II, 331–416)

ITALO CALVINO
(1923—)

Crystals

If the substances that made up the terrestrial globe in its incandescent state had had at their disposal a period of time long enough to allow them to grow cold and also sufficient freedom of movement, each of them would have become separated from the others in a single, enormous crystal.

It could have been different, I know,—*Qfwfq remarked,*—you're telling me: I believed so firmly in that world of crystal that was supposed to come forth that I can't resign myself to living still in this world, amorphous and crumbling and gummy, which has been our lot, instead. I run all the time like everybody else, I take the train each morning (I live in New Jersey) to slip into the cluster of prisms I see emerging beyond the Hudson, with its sharp cusps: I spend my days there, going up and down the horizontal and vertical axes that crisscross that compact solid, or along the obligatory routes that graze its sides and its edges. But I don't fall into the trap: I know they're making me run among smooth transparent walls and between symmetrical angles so I'll believe I'm inside a crystal, so I'll recognize a regular form there, a rotation axis, a constant in the dihedrons, whereas none of all this exists. The contrary exists: glass, those are glass solids that flank the streets, not crystal, it's a paste of haphazard molecules which has invaded and cemented the world, a layer of suddenly chilled lava, stiffened into forms imposed from the outside, whereas inside it's magma just as in the Earth's incandescent days.

I don't pine for them surely, those days: I feel discontented with things as they are, but if, for that reason, you expect me to remember the past with nostalgia, you're mistaken. It was horrible, the Earth without any crust, an eternal incandescent winter, a mineral bog, with black swirls of iron and nickel that dripped down from every crack toward the center of the globe, and jets of mercury that gushed up in high spurts. We made our way through a boiling haze, Vug and I, and we could never manage to touch a solid point. A barrier of liquid rocks that we found before us would suddenly evaporate in our path, disintegrating into an acid cloud; we would rush to pass it, but already we could feel it condensing and striking us like a storm of metallic rain, swelling the thick waves of an aluminum ocean. The substance of things changed around us every minute; the atoms, that is, passed from one state of disorder to another state of disorder and then another still: or rather, practically speaking, everything remained always the same. The only real change would have been the atoms' arranging themselves in some sort of order: this is what Vug and I were looking for, moving in the mixture of the elements without any points of reference, without a before or an after.

Now the situation is different, I admit: I have a wrist watch, I compare the angle of its hands with the angle of all the hands I see; I have an engagement book where the hours of my business appointments are marked down; I have a checkbook on whose stubs I add and subtract numbers. At Penn Station I get off the train, I take the subway, I stand and grasp the strap with one hand to keep my balance while I hold my newspaper up in the other, folded so I can glance over the figures of the stock market quotations: I play the game, in other words, the game of pretending there's an order in the dust, a regularity in the system, or an interpenetration of different systems, incongruous but still measurable, so that every graininess of disorder coincides with the faceting of an order which promptly crumbles.

Before it was worse, of course. The world was a solution of substances where everything was dissolved into everything and the solvent of everything. Vug and I kept on getting lost in its midst, losing our lost places, where we had been lost always, without any idea of what we could have found (or of what could have found us) so as to be lost no more.

We realized it all of a sudden. Vug said: "There!"

She was pointing, in the midst of a lava flow, at something that was taking form. It was a solid with regular, smooth facets and sharp corners; and these facets and corners were slowly expanding, as if at the expense of the surrounding matter, and also the form of the solid was changing, while still maintaining symmetrical proportions. . . . And it wasn't only the form that was distinct from all the rest: it was also the way the light entered inside, passing through it and refracted by it. Vug said: "They shine! Lots of them!"

It wasn't the only one, in fact. On the incandescent expanse where once only ephemeral gas bubbles had risen, expelled from the Earth's bowels, cubes were now coming to the surface and octahedrons, prisms, figures so transparent they seemed airy, empty inside, but instead, as we soon saw, they concentrated in themselves an incredible compactness and hardness. The sparkle of this angled blossoming was invading the Earth, and Vug said: "It's a spring!" I kissed her.

Now you can understand me: if I love order, it's not—as with so many others—the mark of a character subjected to an inner discipline, a repression of the instincts. In me the idea of an absolutely regular world, symmetrical and methodical, is associated with that first impulse and burgeoning of nature, that amorous tension—what you call eros—while all the rest of your images, those that according to you associate passion with disorder, love with intemperate overflow—river fire whirlpool volcano—for me are memories of nothingness and listlessness and boredom.

It was a mistake on my part, it didn't take me long to understand that. Here we are at the point of arrival: Vug is lost; of the diamond eros only dust remains; the simulated crystal that imprisons me now is base glass. I follow the arrows on the asphalt, I line up at the traffic light, and I start again (today I came into New York by car) when the green comes on (as I do every Wednesday because I take) shifting into first

(Dorothy to her psychoanalyst), I try to maintain a steady speed which allows me to pass all the green lights on Second Avenue. This, which you call order, is a threadbare patch over disintegration; I found a parking space but in two hours I'll have to go down again to put another coin in the meter; if I forget they'll tow my car away.

I dreamed of a world of crystal, in those days: I didn't dream it, I saw it, an indestructible frozen springtime of quartz. Polyhedrons grew up, tall as mountains, diaphanous: the shadow of the person beyond pierced through their thickness. "Vug, it's you!" To reach her I flung myself against walls smooth as mirrors; I slipped back; I clutched the edges, wounding myself; I ran along treacherous perimeters, and at every turn there was a different light—diffused, milky, opaque—that the mountain contained.

"Where are you?"

"In the woods!"

The silver crystals were filiform trees, with branches at every right angle. Skeletal fronds of tin and of lead thickened the forest in a geometric vegetation.

In the middle there was Vug, running. "Qfwfq! It's different over there!" she cried. "Gold, green blue!"

A valley of beryllium opened out, surrounded by ridges of every color, from aquamarine to emerald. I followed Vug with my spirit torn between happiness and fear: happiness at seeing how every substance that made up the world was finding its definitive and solid form, and a still vague fear that this triumph of order in such various fashions might reproduce on another scale the disorder we had barely left behind us. A total crystal I dreamed, a topaz world that would leave out nothing: I was impatient for our Earth to detach itself from the wheel of gas and dust in which all the celestial bodies were whirling, ours should be the first to escape that useless dispersal which is the universe.

Of course, if he chooses, a person can also take it into his head to find an order in the stars, the galaxies, an order in the lighted windows of the empty skyscrapers where between nine and midnight the cleaning women wax the floors of the offices. Rationalize, that's the big task: rationalize if you don't want everything to come apart. Tonight we're dining in town, in a restaurant on the terrace of a twenty-fourth floor. It's a business dinner: there are six of us; there is also Dorothy, and the wife of Dick Bemberg. I eat some oysters, I look at a star that's called (if I have the right one) Betelgeuse. We make conversation: we husbands talk about production; the ladies, about consumption. Anyway, seeing the firmament is difficult: the lights of Manhattan spread out a halo that becomes mixed with the luminosity of the sky.

The wonder of crystals is the network of atoms that is constantly repeated: this is what Vug wouldn't understand. What she liked—I quickly realized—was to discover in crystals some differences, even minimal ones, irregularities, flaws.

"But what does one atom out of place matter to you, an exfoliation that's a bit crooked," I said, "in a solid that's destined to be enlarged infinitely according to a regular pattern? It's a single crystal we're working toward, the gigantic crystal. . . ."

"I like them when there're lots of little ones," she said. To contradict me, surely; but also because it was true that crystals were popping up by the thousands at the same time and were interpenetrating one another, arresting their growth where they came in contact, and they never succeeded in taking over entirely the liquid rock from which they received their form: the world wasn't tending to be composed into an ever-simpler figure but was clotting in a vitreous mass from which prisms and octahedrons and cubes seemed to be struggling to be free, to draw all the matter to themselves. . . .

A crater exploded: a cascade of diamonds spread out.

"Look! Aren't they big?" Vug exclaimed.

On every side there were erupting volcanoes: a continent of diamond refracted the sun's light in a mosaic of rainbow chips.

"Didn't you say the smaller they are the more you like them?" I reminded her.

"No! Those enormous ones—I want them!" and she darted off.

"There are still bigger ones," I said, pointing above us. The sparkle was blinding: I could already see a mountain-diamond, a faceted and iridescent chain, a gem-plateau, a Koh-i-noor-Himalaya.

"What can I do with them? I like the ones that can be picked up. I want to have them!" and in Vug there was already the frenzy of possession.

"The diamond will have us, instead. It's the stronger," I said.

I was mistaken, as usual: the diamond was had, not by us. When I walk past Tiffany's, I stop to look at the windows, I contemplate the diamond prisoners, shards of our lost kingdom. They lie in velvet coffins, chained with silver and platinum; with my imagination and my memory I enlarge them, I give them again the gigantic dimensions of fortress, garden, lake, I imagine Vug's pale blue shadow mirrored there. I'm not imagining it: it really is Vug who now advances among the diamonds. I turn: it's the girl looking into the window over my shoulder, from beneath the hair falling across her forehead.

"Vug!" I say. "Our diamonds!"

She laughs.

"Is it really you?" I ask. "What's your name?"

She gives me her telephone number.

We are among slabs of glass: I live in simulated order, I would like to say to her, I have an office on the East Side, I live in New Jersey, for the weekend Dorothy has invited the Bembergs, against simulated order simulated disorder is impotent, diamond would be necessary, not for us to possess it but for it to possess us, the free diamond in which Vug and I were free. . . .

"I'll call you," I say to her, only out of the desire to resume my arguing with her.

In an aluminum crystal, where chance scatters some chrome atoms, the transparency is colored a dark red: so the rubies flowered beneath our footsteps.

"You see?" Vug said. "Aren't they beautiful?"

We couldn't walk through a valley of rubies without starting to quarrel again.

"Yes," I said, "because the regularity of the hexagon . . ."

"Uff!" she said. "Would they be rubies without the intrusion of extraneous atoms? Answer me that!"

I became angry. More beautiful? Or less beautiful? We could go on arguing to infinity, but the only sure fact was that the Earth was moving in the direction of Vug's preferences. Vug's world was in the fissures, the cracks where lava rises, dissolving the rock and mixing the minerals in unpredictable concretions. Seeing her caress walls of granite, I regretted what had been lost in that rock, the exactness of the feldspars, the micas, the quartzes. Vug seemed to take pleasure only in noting how minutely variegated the face of the world appeared. How could we understand each other? For me all that mattered was homogeneous growth, indiscerptibility, achieved serenity; for her, everything had to be separation and mixture, one or the other, or both at once. Even the two of us had to take on an aspect (we still posessed neither form nor future): I imagined a slow uniform expansion, following the crystals' example, until the me-crystal would have interpenetrated and fused with the her-crystal and perhaps together we would have become a unity within the world-crystal; she already seemed to know that the law of living matter would be infinite separating and rejoining. Was it Vug, then, who was right?

It's Monday; I telephone her. It's almost summer already. We spend a day together, on Staten Island, lying on the beach. Vug watches the grains of sand trickle through her fingers.

"All these tiny crystals . . ." she says.

The shattered world that surrounds us is, for her, still the world of the past, the one we expected to be born from the incandescent world. To be sure, the crystals still give the world form, breaking up, being reduced to almost imperceptible fragments rolled by the waves, encrusted with all the elements dissolved in the sea which kneads them together again in steep cliffs, in sandstone reefs, a hundred times dissolved and recomposed, in schists, slates, marbles of glabrous whiteness, simulacra of what they once could have been and now can never be.

And again I am gripped by my stubbornness as I was when it began to be clear that the game was lost, that the Earth's crust was becoming a congeries of disparate forms, and I didn't want to resign myself, and at every irregularity in the porphyry that Vug happily pointed out to me, at every vitrescence that emerged from the basalt, I wanted to persuade myself that these were only apparent flaws, that they were all part of a much vaster regular structure, in which every asymmetry we thought we observed really corresponded to a network of symmetries so complicated we couldn't comprehend it, and I tried to calculate how many billions of sides and dihedral corners this labyrinthine crystal must have, this hypercrystal that included within itself crystals and noncrystals.

Vug has brought a little transistor radio along to the beach with her.

"Everything comes from crystal," I say, "even the music we're hearing." But I know full well that the transistor's crystal is imperfect, flawed, veined with impurities, with rents in the warp of the atoms.

She says: "It's an obsession with you." And it is our old quarrel, continuing. She wants to make me admit that real order carries impurity within itself, destruction.

The boat lands at the Battery, it is evening; in the illuminated network of the skyscraper-prisms I now look only at the dark rips, the gaps. I see Vug home; I go up with her. She lives downtown, she has a photography studio. As I look around I see nothing but perturbations of the order of the atoms: luminescent tubes, TV, the condensing of tiny silver crystals on the photographic plates. I open the icebox, I take out the ice for our whisky. From the transistor comes the sound of a saxophone. The crystal which has succeeded in becoming the world, in making the world transparent to itself, in refracting it into infinite spectral images, is not mine: it is a corroded crystal, stained, mixed. The victory of the crystals (and of Vug) has been the same thing as their defeat (and mine). I'll wait now till the Thelonius Monk record ends, then I'll tell her.

LOREN EISELEY
(1907—)

Notes of an Alchemist

Crystals grow
      under fantastic pressures in the deep
            crevices and confines of
the earth.
      They grow by fires,
           by water trickling slowly
in strange solutions
      from the walls of caverns.
They form
        in cubes, rectangles,
         tetrahedrons,
            they may have
their own peculiar axes and
        molecular arrangements
           but they,
    like life,
        like men,
          are twisted by
the places into which
      they come.

I have only
    to lift my hands
        to see
the acid scars of old encounters.
          In my brain

as in the brains of all mankind
                                distortions riot
and the serene
                        quartz crystal of tomorrow is
most often marred
                        by black ingredients
                                        caught blindly up,
but still
                        no one knows surely why
                                        specific crystals meet
in a specific order.
                        Therefore we grasp
two things:
                        that rarely
two slightly different substances will grow
                                        even together
but the one added ingredient
                        will transfigure
a colorless transparency
                        to midnight blue
or build the rubies' fire.
                        Further, we know
that if one grows a crystal
                        it should lie
under the spell of its own fluid
                                be
                        kept in a cool cavern
                        remote
                        from any violence or
                                intrusion from the dust.

So we
        our wise men
                in their wildernesses
                                have sought
to charm to similar translucence
                the cloudy crystal of the mind.
We must then understand
                        that order strives

against the unmitigated chaos lurking
along the convulsive backbone of the world.
Sometimes I think that we
                    in varying degrees are grown
like the wild crystal,
                    now inert,
now flashing red,
            but I
                    within my surging molecules
by nature cling
            to that deep sapphire blue
                        that marks the mind of one
long isolate
            who knows and does reflect
starred space and midnight,
                    who conceives therefore
that out of order and disorder
                        perpetually clashing and reclashing
come the worlds.
            Thus stands my study from the vials and furnaces
of universal earth. I leave it here
                    for Heracleitus
if he comes again
            in the returnings of the Giant Year.

JOHN UPDIKE

(1932—)

From *Midpoint*

III. THE DANCE OF THE SOLIDS

Argument: In stanzas associated with allegory the actual
atomic structure of solids unfolds. Metals, Ceramics, and
Polymers. The conduction of heat, electricity, and light; non-

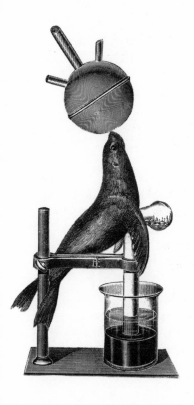

symmetry and magnetism. Solidity emerges as intricate and giddy.

All things are Atoms: Earth and Water, Air
   And Fire, all *Democritus* foretold.
Swiss *Paracelsus*, in's alchemic lair,
   Saw Sulphur, Salt, and Mercury unfold
   Amid Millennial hopes of faking Gold.
Lavoisier dethroned Phlogiston; then
   Molecular Analysis made bold
   Forays into the gasses: Hydrogen
Stood naked in the dazzled sight of Learned Men.

The Solid State, however, kept its grains
   Of Microstructure coarsely veiled until
   X-ray diffraction pierced the Crystal Planes

That roofed the giddy Dance, the taut Quadrille
Where Silicon and Carbon Atoms will
Link Valencies, four-figured, hand in hand
With common Ions and Rare Earths to fill
The Lattices of Matter, Salt or Sand,
With tiny Excitations, quantitatively grand.

The *Metals*, lustrous Monarchs of the Cave,
    Are ductile and conductive and opaque
    Because each Atom generously gave
    Its own Electrons to a mutual Stake,
    A Pool that acts as Bond. The Ions take
The stacking shape of Spheres, and slip and flow
    When pressed or dented; thusly *Metals* make
    A better Paper Clip than a Window,
Are vulnerable to Shear, and, heated, brightly glow.

Ceramic, muddy Queen of Human Arts,
    First served as Simple Stone, Feldspar supplied
    Crude Clay; and Rubies, Porcelain, and Quartz
    Came each to light. Aluminum Oxide
    Is typical—a Metal close-allied
    With Oxygen ionically; no free
    Electrons form a lubricating tide,
    Hence Empresslike, *Ceramics* tend to be
Resistant, pourous, brittle, and refractory.

Prince *Glass*, Ceramic's son, though crystal-clear,
    Is no wise crystalline. The fond Voyeur
    And Narcissist alike devoutly peer
    Into Disorder, the Disorderer
    Being Covalent Bondings that prefer
    Prolonged Viscosity and spread loose nets
    Photons slip through. The average *Polymer*
    Enjoys a Glassy state, but cools, forgets
To slump, and clouds in closely patterned Minuets.

The *Polymers*, those giant Molecules,
    Like Starch and Polyoxymethylene,

Flesh out, as protein serfs and plastic fools,
This Kingdom with Life's Stuff. Our time has seen
The synthesis of Polyisoprene
And many cross-linked Helixes unknown
To *Robert Hooke;* but each primordial Bean
Knew Cellulose by heart. *Nature* alone
Of Collagen and Apatite compounded Bone.

What happens in these Lattices when *Heat*
Transports Vibrations through a solid mass?
$T = 3Nk$[43] is much too neat;
A rigid Crystal's not a fluid Gas.
Debye[44] in 1912 proposed Elas-
Tic Waves called *phonons* that obey Max Planck's[45]
$E = \frac{h}{v}$. Though amorphous Glass,
Umklapp Switchbacks,[46] and Isotopes play pranks
Upon his Formulae, *Debye* deserves warm Thanks.

Electroconductivity depends
On Free Electrons: in Germanium
A touch of Arsenic liberates; in blends
Like Nickel Oxide, *Ohms* thwart Current. From
Pure Copper threads to wads of Chewing Gum
Resistance varies hugely. Cold and Light
As well as "doping"[47] modify the sum
Of *Fermi* levels,[48] Ion scatter, site
Proximity, and other Factors recondite.

Textbooks and Heaven only are Ideal;
Solidity is an imperfect state.

[43]An equation relating the "degrees of freedom" of motion a solid can have based on temperature.
[44]Peter Joseph William Debye (1884–1966), Dutch physicist who succeeded Einstein as professor of theoretical physics at the University of Zurich. After shifting his interest to the thermal movement of atoms in crystals affected by X-ray interferences, he derived formal patterns for the structural analysis of crystals. He won the Nobel Prize in chemistry in 1936.
[45]Planck derived an energy constant, h. This "quantum of action" is the ratio of energy to frequency.
[46]A specific interaction of electron or lattice waves in a solid.
[47]Adding impurities to a semiconductor to achieve a desired characteristic.
[48]Electron levels in a metal.

Within the cracked and dislocated Real
Nonstoichiometric crystals[49] dominate.
Stray Atoms sully and precipitate;
Strange holes, *excitons*, wander loose; because
Of Dangling Bonds, a chemical Substrate
Corrodes and catalyzes—surface Flaws
Help Epitaxial Growth[50] to fix adsorptive claws.

White Sunlight, *Newton* saw, is not so pure;
A Spectrum bared the Rainbow to his view.
Each Element absorbs its signature:
Go add a negative Electron to
Potassium Chloride; it turns deep blue,
As Chromium incarnadines Sapphire.
Wavelengths, absorbed, are reëmitted through
Fluorescence, Phosphorescence, and the higher
Intensities that deadly *Laser Beams* require.

Magnetic Atoms, such as Iron, keep
Unpaired Electrons in the middle shell,
Each one a spinning Magnet that would leap
The *Bloch* Walls[51] wherat antiparallel
Domains converge. Diffuse Material
Becomes *Magnetic* when another Field
Aligns domains like Seaweed in a swell.
How nicely microscopic forces yield,
In Units growing visible, the World we wield!

[49]Asymmetrical structure in which one molecule is either substituted or missing.
[50]Occurs when one crystal grows on the surface of the other.
[51]Transition layers between ferromagnetic domains that are magnetized in different directions.

BIOLOGY

INTRODUCTION

Philosophers more grave than wise
Hunt science down in butterflies;
Or fondly poring on a spider,
Stretch human contemplation wider.
—John Gay

In ancient times biology was almost equivalent to medicine and anatomy. Because of their religious beliefs, the ancient Egyptians required that the body of a deceased person be preserved for resurrection. This led to the practice of embalming, which began around 2500 B.C. and which indirectly involved the study of anatomy. In their art it was necessary for embalmers to remove the internal organs and the brain of the deceased, a practice that facilitated a great deal of specific knowledge about the body. Although there is no surviving papyrus specifically dealing with embalming techniques, there are several that demonstrate the Egyptians' superiority in the medical area.

The Greeks were the first to take up the two branches of biology—zoology and botany—as academic areas of empirical study. While Aristotle was the model of the empiricist, his teacher, Plato, in the *Timaeus*, demonstrates quite a bit of anatomical knowledge—though it is doubtful that he had any practical experience. While he knew where the organs were in the body, his descriptions of their functions leave something to be desired: The purpose of the liver was prophecy, while the spleen was viewed as a napkin to keep the liver clean!

Plato was a theoretician but Aristotle spent much of his adulthood studying plants and animals (all his botanical works are lost) and wrote numerous biological works. He was one of the first to classify animals into groups, but when it came to death he believed that all living things died of the same cause. In his work *On Breathing* he suggests that all natural deaths are caused by a lack of heat; in plants it is withering and in animals it is senility.

When the library at Alexandria was founded, it brought together scientists from all the Greek islands in an atmosphere conducive to study. While further advances in biology were made in Alexandria, particularly in anatomy, it was Pliny the Elder who wrote the book that was to be the most influential work on biology until the Renaissance.

Pliny was interested in everything, but he was not a scientist. His *Natural History* was probably the second best-selling book, after the Bible, for fifteen centuries. In the selection presented here Pliny's awe of the natural world is evident. He is amazed that man in nature is such a fragile physical and moral species. In comparing man's heart to the organ in other species he indulges in fascinating narrative digressions that account for his great vitality as an encyclopedist.

Unlike Pliny, Galen of Pergamum was a scientist. His detailed knowledge of anatomy was probably derived from his practice as a surgeon to the Roman gladiators. Because of the superstitions of his times, he was not permitted to dissect human bodies. His book *On Anatomical Preparations* was the standard text for well over a thousand years. Unaware of the recirculation of the blood, Galen believed blood was used up on its one-way trip throughout the body and new blood was created in the liver. This belief was not corrected until the time of Flemish anatomist Andreas Vesalius (1514–64).

Biology did not fare well during the Middle Ages. The Arabs were more interested in astronomy and chemistry and neglected the study of biology. The Western world was absorbed with Christianity, so science in general was not advanced. It is not until the Renaissance that the study of biology was reawakened.

As the Renaissance was much more liberal in its attitude toward human dissection, the various universities organized public dissections called "anatomies." These were rather curious affairs. There seems to have been a stigma associated with actually dissecting a body, so the professor read from Galen while a technician did the actual cutting. If the evidence contradicted what Galen said, then Galen was accepted as the one to be believed. Further, since these were rather crowded affairs, the students probably could not see much—the educational value of these anatomies is somewhat doubtful.

Two artists who did their own dissections were Leonardo da Vinci and Michelangelo, who needed detailed anatomical knowledge for their art. The one anatomist who did his own dissections was the great Vesalius, who obtained his corpses from executions. Realizing that Galen was frequently in error, he deduced that Galen had gotten much of his knowledge from dissecting apes and not humans. To prove his point, he assembled the skeletons of an ape and a human before a large audience and showed that where the skeletons differed, Galen was wrong in his description of human anatomy.

Such demonstrations as Vesalius' led the way for a break with tradition and opened biology up to new thinking and experimentation. It was this atmosphere that permitted William Harvey to discover the true nature of the human circulatory system. As our selection from his *Circulation of the Blood* shows, he did not totally reject his earlier counterparts but rather was willing to accept what he viewed as good in both Galen and Aristotle. In his *Brief Lives*, Harvey's friend, the amusing anecdotist John Aubrey, presents an intimate sense of Harvey's humanity as well as the impact of his discovery. Abraham Cowley, the official poetic spokesman of the Royal Society, pays tribute to Harvey's "noble" circulation in a classical ode.

It was the invention of the microscope in the seventeenth century that drew biologists away from the study of anatomy and toward the study of "wee beasties." Indeed, the pioneers of microscopy such as Antonie van Leeuwenhoek (1632–1733) and Robert Hooke (1635–1703) created a whole new area of study—microbiology. One of the things that fascinated seventeenty-century biologists was the study of sperm under the microscope. These "animalcules," as they were called, led biologists to speculate on just how small living creatures could be. This gave rise to all kinds of calculations on the size, weight, etc., of sperm. Henry Baker's *Of Microscopes and the Discoveries Made Thereby* is a good example of this preoccupation.

The new world opened by the microscope also prompted speculation outward. Since life had been discovered to exist even on the surface of the skin, wasn't it likely that new life would be discovered on the planets outside our own? This is precisely the suggestion made by Joseph Addison. When he wonders at the diversity of life in *The Spectator*, he raises questions about the origin of this diversity. In time, such questioning led to the theory of evolution.

In Addison's age naturalists were still engaged in Pliny's great quest, to catalog the plenitude of nature. This was expressed for the eighteenth century in the concept of the "great chain of being," the idea that from highest to lowest life there existed an unbroken organic hierarchy. It descended from God through seven orders of angels to man (the center) and from man through the animal kingdom to its most minute members—those same members revealed by the microscope! The function of a naturalist as a life scientist was to confirm proof of this unbroken chain and so help man keep his central place and give praise to his creator.

Based largely on the work of Pliny, Aristotle, and the ancients, great compendia illustrating nature's diversity and order were produced in the Renaissance. They commonly mingled fact and myth, unable to distinguish between them because very few of the authors had access to the animals they wrote about. How frequently could a European see a sloth, rhinoceros, or whale? Many artists drew the more exotic animals from their imagination or from specimens brought back by voyagers to remote lands. (The gorilla, for example, was never heard of or seen until the nineteenth-century exploration of Africa.) Some animals, like the sloth, remained fixed to their medieval, symbolic value. Sloth was a deadly sin; it was also a South American and difficult to come by! It was an animal that Charles Darwin would get to see in his travels on the *Beagle*.

Before Darwin's far-reaching scientific voyage, most of the

ground-breaking naturalists worked closer to home. Izaak Walton's seventeenth-century compendium *The Compleat Angler* is the work of a man with a passion for fish, and in particular the fish of his native English streams. The Reverend Gilbert White in the eighteenth century had a similar passion for local British birds. He observed their more intimate life habits on his daily walks and recorded his observations with detailed accuracy.

In the eighteenth century naturalists working from observation started to specialize. Not only were they attempting to catalog more completely the species and subspecies of particular genera, they also became more interested in the mechanisms that made life tick. As already mentioned, sexuality was one of the most fascinating mechanisms made more comprehensible by the development of the microscope. The Swedish botanist Carl Linnaeus (1707–78) used sexuality as the basis for ordering the plant kingdom. His system, popularized in Erasmus Darwin's poetic interpretation, "Loves of the Plants," anthropomorphized the plants and their "love" relationships. Stamen and pistils, male and female organs, became swains and nymphs, and frequently Darwin had to invent most peculiar courtship rituals to account for one female and six males! It is easy to see how only the fittest might survive.

Evolutionary theory did not arise *ex nihilo*. As early as the seventeenth century natural scientists were beginning to discuss the nature of fossils and how life left the sea and moved onto the land. The Reverend White, whose native Selborne was surrounded by chalk cliffs, could not help connecting the fossils found there with the observable diversity of species. Such connections would be necessary for the theory of evolution. It was a theory that could not come from a laboratory but would have to originate from someone wanting to connect life in all its forms with the evidence of a past. The obvious question that had to be asked was "How is all the diversity possible?" It would seem as if such differences would indicate the working of chance, but one could not ignore the many similarities as well. How would one reconcile the two?

The answer was put forward in the 1850s by two natu-

ralists working independently of each other—Charles Darwin
and Alfred Russel Wallace. Darwin had dropped out of medical school after two years to study for the clergy at Cambridge. There he had become interested in collecting beetles.
When in 1831 the Admiralty sent out the survey ship *Beagle* to
map the coast of South America, Darwin had signed on as
ship's naturalist. After five years with the ship, Darwin had
returned to England and had begun thinking about the diversity of species. Convinced that species are not immutable, he
believed that when separated they could change in different
directions. This, of course, was anti-biblical. Two years later,
Darwin hit upon the answer science was seeking. He was
reading Robert Thomas Malthus's *Essay on Population*, and
when Malthus described how population increases faster than
food supply Darwin realized that there must be competition
for the food, with the weaker losing the competition and
dying and the stronger surviving and forming new species.

Darwin was not anxious to publish his theory, knowing it
would be highly controversial. It wasn't until 1842 that he
wrote a 53 page draft of it. Two years later, he had a 230-page
manuscript that he gave to his wife with instructions to publish
it in the event of his death. The manuscript lay unattended
until another man—Wallace—hit upon the same theory and
forced Darwin to act.

Unlike Darwin, Wallace had not had a wealthy father to
support him, so he had been forced to leave school at an early
age. He had educated himself and gone to school at night
while earning his living as a surveyor—an occupation that
had put him in close contact with nature. He had made
friends with Henry Bates, a naturalist, and soon become interested in collecting specimens of all kinds. In 1848 he and Bates
had set off for the Amazon jungle to earn livings collecting
specimens. While they were in the jungle, the crucial question
arose in Wallace's mind: How had all this diversity of nature
come about? He spent four years in the Amazon and then
packed his specimens and sailed for home. Unfortunately, his
specimens were all destroyed when his ship caught fire.

In 1854 Wallace once again set out to collect specimens,
this time in the Malay Archipelago, where he remained for

eight years. While there he too read Malthus, and precisely the same question *and* answer struck him as they had Darwin over a decade earlier. Unlike Darwin, Wallace was quick to write down his theory and, knowing that Darwin was interested in such matters, he sent a copy of his paper to him. The year was 1858.

Darwin did not know what to do: Wallace's paper was a perfect abstract of Darwin's longer thesis. Friends who were familiar with Darwin's work settled the issue. Wallace's paper and one by Darwin were read in the absence of both authors at a meeting of the Linnaean Society. The papers did not cause much of a stir at the meeting, but their reading did prompt Darwin to publish *The Origin of Species* in 1859. The book was an instant sensation. It marshaled all the facts and logic necessary to support a growing belief; that the living world was changing; that it had not merely been created and left in the same state since the beginning; that it and the species that inhabit it were in flux.

The implications of flux caused tremendous disturbance among Darwin's contemporaries. His notion of a struggle for survival, a fang-and-claw combat up the evolutionary scale, interfered with Christian notions of morality. The poet laureate Alfred, Lord Tennyson, forced to consider God and nature at strife, could only express a glimmer of hope that "somehow good" would be the "final goal" of all this "ill." In the meantime his cynical imagination lead to a vision of harsh nature condemning man—like so many species before him—to extinction. Tennyson struggled for many years to reconcile himself and his audience to the depressing future for mankind implied by evolution theory.

Only later were writers more able to respond to the comedic implications of organic flux. Samuel Butler's satiric utopia, *Erewhon*, presents a parody of evolution applied to machines; it probably seems less absurd to the reading audience of the computer age than it did to his own.

Whether mechanical or animate, a new species popping up was an unthinkable violation of the religious position asserting God's six-day creation of life in its finished form. It was more directly unthinkable that man, who was the central link in the "chain of being" and the favored species created in God's own image, descended from the same ancestral primate that also gave rise to apes. Yet a scant century after Darwin, Desmond Morris—formerly the director of the London Zoo—was highly acclaimed for his study of man as "the naked ape," a study summarized in John Updike's comic poem putting man in his evolutionary place.

In the latter part of the nineteenth century and throughout the twentieth other important biological discoveries reinforced the notion of man's tenuous position in life. It was realized that diseases were caused by minute organisms that lived in their hosts. This caused great concern that mankind was at the mercy of creatures that could not even be seen. In a paper read in 1878 Louis Pasteur assures the public that inoculations will enable science to control microbes.

The ambivalence of the single cell, its potential danger but also its profound importance as the smallest meaningful unit of life, has come to have particular significance in our time. Man's disinheritance as a favorite species has to some degree been offset by romantic individualism. The cell is the ultimate individual, and writers have taken any number of microbes as metaphors for particular aspects of individuality. Updike's amoeba is a microcosm of the engulfing world that ingests the frightened ego. Muriel Rukeyser's paramecium explains love as a sharing of nuclear matter. Richard Eberhart's cancer cells are "originals of imagination," deviant forms as deadly and beautiful as art itself. Under the microscope Miroslav Holub discovers "a map of the universe."

The invention of the electron microscope permitted an even closer scrutiny of life's mechanisms. Genetics became the dominant theme of these investigations and culminated in the unraveling of the molecular structure of DNA in 1953 by James Watson (1928—) and Francis Crick (1916—). The twisting double helix is compared both to Marcel Duchamp's "Nude Descending a Staircase" and to a transcendent resurrection in a poem by May Swenson. Biology reveals a staircase spiraling through life and death. For some writers, like Kathleen Raine and Nathaniel Tarn, biology opens on philosophy. For Zbigniew Herbert, the Polish poet who imagines the biology teacher destroyed in the Holocaust, faith in the beetle is perhaps easier than faith in man. In nature, in the woods, he finds cause for compassion. The lesson of biology becomes for him, as for Kenneth Rexroth, a lesson of love that makes science human.

PLATO
(428–348 B.C.)

From *Timaeus*

As I said at first, when all things were in disorder, God created in each thing in relation to itself, and in all things in relation to each other, all the measures and harmonies which they could possibly receive. For in those days nothing had any proportion except by accident, nor was there anything deserving to be called by the names which we now use—as, for example, fire, water and the rest of the elements. All these the creator first set in order, and out of them he constructed the universe, which was a single animal comprehending in itself all other animals, mortal and immortal. Now of the divine, he himself was the creator, but the creation of the mortal he committed to his offspring. And they, imitating him, received from him the immortal principle of the soul, and around this they proceeded to fashion a mortal body, and made it to be the vehicle of the soul, and constructed within the body a soul of another nature which was mortal, subject to terrible and irresistible affections—first of all, pleasure, the greatest incitement to evil; then, pain, which deters from good; also rashness and fear, two foolish counselors, anger hard to be appeased, and hope easily led astray—these they mingled with irrational sense and with all-daring love according to necessary laws, and so framed man. Wherefore, fearing to pollute the divine any more than was absolutely unavoidable, they gave to the mortal nature a separate habitation in another part of the body, placing the neck between them to be the isthmus and boundary, which they constructed between the head and breast, to keep them apart. And in the breast, and in what is termed the thorax, they incased the mortal soul, and as the one part of this was superior and the other inferior they divided the cavity of the thorax into two parts, as the women's and men's apartments are divided in houses, and placed the midriff to be a wall of partition between them. That part of the inferior soul which is endowed with courage and passion and loves contention, they settled nearer the head, midway between the midriff and the neck, in order that being obedient to the rule of reason it might join with it in controlling and restraining the desires when they are no longer willing of their own accord to obey the word of command issuing from the citadel.

The heart, the knot of the veins and the fountain of the blood which races through all the limbs, was set in the place of guard, that, when the might of passion was roused by reason making proclamation of any wrong assailing them from without or being perpetrated by the desires within, quickly the whole power of feeling in the body, perceiving these commands and threats, might obey and follow through every turn and alley, and thus allow the principle of the best to have the command in all of them. But the gods, foreknowing that the palpitation of the heart in the expectation of danger and excitement of passion must cause it to swell and become inflamed, formed and implanted as a supporter to the heart the lung, which was, in the first place, soft and bloodless, and also had within hollows like the pores of a sponge, in order that by receiving the breath and the drink, it might give coolness and the power of respiration and alleviate the heat. Wherefore they cut the air channels leading to the lung, and placed the lung about the heart as a soft spring, that, when passion was rife within, the heart, beating against a yielding body, might be cooled and suffer less, and might thus become more ready to join with passion in the service of reason.

The part of the soul which desires meats and drinks and the other things of which it has need by reason of the bodily nature, they placed between the midriff and the boundary of the navel, contriving in all this region a sort of manager for the food of the body, and there they bound it down like a wild animal which was chained up with man, and must be nourished if man was to exist. They appointed this lower creation his place here in order that he might be always feeding at the manger, and have his dwelling as far as might be from the council chamber, making as little noise and disturbance as possible, and permitting the best part to advise quietly for the good of the whole and the individual. And knowing that this lower principle in man would not comprehend reason, and even if attaining to some degree of perception would never naturally care for rational notions, but that it would be especially led by phantoms and visions night and day—planning to make this very weakness serve a purpose, God combined with it the liver and placed it in the house of the lower nature, contriving that it should be solid and smooth, and bright and sweet, and should also have a bitter quality in order that the power of thought, which proceeds from the mind, might be reflected as in a mirror which receives likenesses of objects and gives back images of them to the sight, and so might strike terror into the desires when, making use of the bitter part of the liver, to which it is akin, it comes threatening and invading, and diffusing this bitter element swiftly through the whole liver produces colors like bile, and contracting every part makes it wrinkled and rough, and twisting out of its right place and contorting the lobe and closing and shutting up the vessels and gates causes pain and loathing. And the converse happens when some gentle inspiration of the understanding pictures images of an opposite character, and allays the bile and bitterness by refusing to stir or touch the nature opposed to itself, but by making use of the natural sweetness of the liver corrects all things and makes them to be right and smooth and free, and renders the portion of the soul which resides about the liver happy and joyful, enabling it to pass the night in peace, and to practice divination in sleep, inasmuch as it has no share in mind and reason. For the authors of our being, remembering the command of their father when he bade them create the human race as good as they could, that they might correct our inferior parts and make them to attain a measure of truth, placed in the liver the seat of divination. And herein is a proof that God has given the art of divination not to the wisdom, but to the foolishness of man. No man, when in his wits, attains prophetic truth and inspiration, but when he receives the inspired word, either his intelligence is enthralled in sleep or he is demented by some distemper or possession. And he who would understand what he remembers to have been said, whether in a dream or when he was awake, by the prophetic and inspired nature, or would determine by reason the meaning of the apparitions which he has seen, and what indications they afford to this man or that, of past, present, or future good and evil, must first recover his wits. But, while he continues demented, he cannot judge of the visions which he sees or the words which he utters: the ancient saying is very true—that "only a man who has his wits can act or judge about himself and his own affairs." And for this reason it is customary to appoint interpreters to be judged of the true inspiration. Some persons call them prophets, being blind to the fact that they are only the expositors of dark sayings and visions, and are not to be called prophets at all, but only interpreters of prophecy.

Such is the nature of the liver, which is placed as we have described in order that it may give prophetic intimations. During the life of each individual these intimations are plainer, but after his death the liver becomes blind and delivers oracles too obscure to be intelligible. The neighboring organ [the spleen] is situated on the left-hand side and is constructed with a view of keeping the liver bright and pure—like a napkin, always ready-prepared and at hand to clean the mirror. And hence, when any impurities arise in the region of the liver by reason of disorders of the body, the loose nature of the spleen, which is composed of a hollow and bloodless tissue, receives them all and clears them away, and when filled with the unclean matter, it swells and festers, but, again, when the body is purged, shrinks and settles down into the same place as before.

ARISTOTLE

(384–322 B.C.)

From *On Breathing*

CHAPTER XVII

Generation and death, therefore, are common to all animals; but the modes are specifically different. For corruption is not a thing indifferent, but possesses something common. And death is either violent, indeed, when the principle of it is externally derived; but according to nature, when the principle of it is internal (i.e., when it arises from heat consuming the moisture). And the composition of any part of the animal is from such a principle, and is not any adventitious passion. In plants, therefore (the consumption of the moisture by which their natural death is produced), it is called dryness; but in animals this consumption is denominated old age. Death and corruption, however, subsist similarly in all imperfect living beings; but in such as are perfect, their mode of subsistence is nearly similar, yet differs in some respects. But I call imperfect living beings, such as eggs, and such seeds of plants as have not taken root. Corruption, therefore, is produced, in all living beings from a defect of heat; but in perfect animals it is produced in that part in which the principle of their essence is contained. And this part, as we have before observed, is that in which the superior and inferior part are conjoined; and which in plants is the middle of the blossom and the root, but in sanguineous animals is the heart, and in the exsanguious, that which is analogous to the heart. Of these, however, some have many principles in capacity, yet not in energy. Hence, also, some insects live when divided; and of sanguineous animals, such as are not very vital, or tortoises, live for a long time, when deprived of the heart, and then walk, while they are yet of a small size, because their nature is not well composed. The like, also, takes place in insects. But the principle of life fails in those that possess it, when the heat of which it participates is not refrigerated; for, as we have frequently observed, it is consumed itself by itself. When, therefore, the lungs in some animals, and the gills in others, become hard and terrestrial through length of time, they are unable to move these parts, and can neither elevate nor contract them. At length, however, intension taking place, the fire is consumed. Hence, also, from small passions and (diseases) acceding in old age, they rapidly die; for because the heat is but little, or being, for the most part, evaporated in the length of life, if any intension is produced of a part, it is swiftly extinguished; just as a momentary, gentle wind, is capable of extinguishing a slender flame. Hence, also, death in old age is unattended with pain. For old men die without any violent passion happening to them; and the liberation of the soul is entirely without sensation. And in diseases in which the lungs become hard, either from tubercles, or excrementitious matter, or diseased heat such as that of a fever, the sick frequently respire, because they cannot much raise and depress the lungs. But at length, when the lungs can no longer be moved, the animal expiring, dies.

CHAPTER XVIII

Generation, therefore, is the first participation, in conjunction with heat, of the nutritive soul. But life is the permanency of this soul. Youth is the increase of the first refrigerative part; but old age is the decrease of this said. And acme is a medium between these. And with respect to death and corruption, when they are violent, indeed, they are caused by the extinction and consumption of heat; for it may be corrupted from both these causes. But death and corruption, when they are natural, arise from the consumption of vital heat, produced by the length and perfection of time. In plants, therefore, this consumption is called dryness, but in animals, death. But death, in old age, is the consumption of vital heat, from an imbicility of refrigerating, arising from old age. And thus we have shown what generation, life and death, are, and from what causes they are inherent in animals.

PLINY THE ELDER

(23–79)

From *Natural History*

BOOK VII

MAN, HIS BIRTH, HIS ORGANIZATION, AND THE INVENTION
OF THE ARTS

Chapter 1—Man

Such then is the present state of the world, and of the countries, nations, more remarkable seas, islands, and cities which it contains. The nature of the animated beings which exist upon it, is hardly in any degree less worthy of our contemplation than its other features; if, indeed, the human mind is able to embrace the whole of so diversified a subject. Our first attention is justly due to Man, for whose sake all other things appear to have been produced by Nature; though, on the other hand, with so great and so severe penalties for the enjoyment of her bounteous gifts, that it is far from easy to determine, whether she has proved to him a kind parent, or a merciless step-mother.

In the first place, she obliges him alone, of all animated beings, to clothe himself with the spoils of the others; while, to the rest, she has given various kinds of coverings, such as shells, crusts, spines, hides, furs, bristles, hair, down, feathers, scales, and fleeces. The very trunks of the trees even, she has protected against the effects of heat and cold by a bark, which is, in some cases, twofold. Man alone, at the very moment of his birth cast naked upon the naked earth, does she abandon to cries, to lamentations and, a thing that is the case with no other animal whatever, to tears: this, too, from the very moment that he enters upon existence. But as for laughter, why, by Hercules!—to laugh, if but for an instant only, has never been granted to man before the fortieth day from his birth, and then it is looked upon as a miracle of precocity. Introduced thus to the light, man has fetters and swathings instantly put upon all his limbs, a thing that falls to the lot of none of the brutes even that are born among us. Born to such singular good fortune, there lies the animal, which is destined to command all the others, lies, fast bound hand and foot, and weeping aloud! such being the penalty which he has to pay on beginning life, and that for the sole fault of having been born. Alas! for the folly of those who can think after such a beginning as this, that they have been born for the display of vanity!

The earliest presage of future strength, the earliest bounty of time, confers upon him nought but the resemblance to a quadruped. How soon does man gain the power of walking? How soon does he gain the faculty of speech? How soon is his mouth fitted for mastication? How long are the pulsations of the crown of his head to proclaim him the weakest of all animated beings? And then, the diseases to which he is subject, the numerous remedies which he is obliged to devise against his maladies, and those thwarted every now and then by new forms and features of disease. While other animals have an instinctive knowledge of their natural powers; some, of their swiftness of pace, some of their rapidity of flight, and some again of their power of swimming; man is the only one that knows nothing, that can learn nothing without being taught; he can neither speak, nor walk, nor eat, and, in short, he can do nothing, at the prompting of nature only, but weep. For this it is, that many have been of opinion, that it were better not to have been born, or if born, to have been annihilated at the earliest possible moment.

To man alone, of all animated beings, has it been given, to grieve, to him alone to be guilty of luxury and excess; and that in modes innumerable, and in every part of his body. Man is the only being that is a prey to ambition, to avarice, to an immoderate desire of life, to superstition,—he is the only one that troubles himself about his burial, and even what is to become of him after death. By none is life held on a tenure more frail; none are sensible of fears more bewildering; none are actuated by rage more frantic and violent. Other animals, in fine, live at peace with those of their own kind; we only see them unite to make a stand against those of a different species. The fierceness of the lion is not expended in fighting with its own kind; the sting of the serpent is not aimed at the serpent; and the monsters of the sea even, and the fishes, vent their rage only on those of a different species. But with man,—by Hercules! most of *his* misfortunes are occasioned by man.

*   *   *

BOOK XI   INTRODUCTION

Chapter 69—The Heart; the Blood; the Vital Spirit

In all other animals but man the heart is situate in the middle of the breast; in man alone it is placed just below the pap on the left-hand side, the smaller end terminating in a point, and bearing outward. It is among the fish only that this point is turned towards the mouth. It is asserted that the heart is the first among the viscera that is formed in the foetus, then the brain, and last of all, the eyes: it is said, too, that the eyes are the first organs that die, and the heart the very last of all. The heart also is the principal seat of the heat of the body; it is constantly palpitating, and moves as though it were one animal enclosed within another. It is also enveloped in a membrane equally supple and strong, and is protected by the bulwarks formed by the ribs and the bone of the breast, as being the primary source and origin of life. It contains within itself the primary receptacles for the spirit and the blood, in its sinuous cavity, which in the larger animals is threefold, and in all twofold at least: here it is that the veins, which branch out into the fore-part and the back of the body, and which, spreading out in a series of branches, convey the vital blood by other smaller veins over all parts of the body. This is the only one among the viscera that is not affected by maladies, nor is it subject to the ordinary penalties of human life; but when injured it produces instant death. While all the other viscera are injured, vitality may still remain in the heart.

Chapter 70—Those Animals Which Have the Largest Heart, and Those Which Have the Smallest. What Animals Have Two Hearts.

Those animals are looked upon as stupid and lumpish which have a hard, rigid heart, while those in which it is small are courageous, and those are timid which have it very large. The heart is the largest, in proportion to the body, in the mouse, the hare, the ass; the stag, the panther, the weasel, the hyæna, and all the animals, in fact, which are timid, or dangerous only from the effects of fear. In Paphlagonia the partridge has a double heart. In the heart of the horse and the ox there are bones sometimes found. It is said that the heart increases every year in man, and that two drachmæ in weight are added yearly up to the fiftieth year, after which period it decreases yearly in a similar ratio; and that it is for this reason that men do not live beyond their hundredth year, the heart then failing them: this is the notion entertained by the Egyptians, whose custom it is to embalm the bodies of the dead, and so preserve them. It is said that men have been born with the heart covered with hair, and that such persons are excelled by none in valour and energy; such, for instance, as Aristomenes, the Messenian, who slew three hundred Lacedæmonians. Being covered with wounds, and taken prisoner, he, on one occasion, made his escape by a narrow hole which he discovered in the stone quarry where he was imprisoned, while in pursuit of a fox which had found that mode of exit. Being again taken prisoner, while his guards were fast asleep he rolled himself towards a fire close by, and at the expense of his body, burnt off the cords by which he was bound. On being taken a third time, the Lacedæmonians opened his breast while he was still alive, and his heart was found covered with hair.

GALEN OF PERGAMUM

(131–200)

From *On the Natural Faculties*

BOOK III

14. And further, it has been shown in other treatises that all the arteries possess a power which derives from the heart, and by virtue of which they dilate and contract.

Put together, therefore, the two facts—that the arteries have this motion, and that everything, when it dilates, draws neighbouring matter into itself—and you will find nothing strange in the fact that those arteries which reach the skin draw in the outer air when they dilate, while those which anastomose at any point with the veins attract the thinnest and most vaporous parts of the blood which these contain, and as for those arteries which are near the heart, it is on the heart itself that they exert their traction. For, by virtue of the tendency by which a vacuum becomes refilled, the lightest and thinnest part obeys the tendency before that which is heavier and thicker. Now the lightest and thinnest of anything in the body is firstly pneuma, secondly vapour, and in the third place that part of the blood which has been accurately elaborated and refined.

These, then, are what the arteries draw into themselves on every side; those arteries which reach the skin draw in the outer air (this being near them and one of the lightest of things); as to the other arteries, those which pass up from the heart into the neck, and that which lies along the spine, as also such arteries as are near these—draw mostly from the heart itself; and those which are farther from the heart and skin necessarily draw the lightest part of the blood out of the veins. So also the traction exercised by the diastole of the arteries which go to the stomach and intestines takes place at the expense of the heart itself and the numerous veins in its neighbourhood; for these arteries cannot get anything worth speaking of from the thick heavy nutriment contained in the intestines and stomach, since they first become filled with lighter elements. For if you let down a tube into a vessel full of water and sand, and suck the air out of the tube with your mouth, the sand cannot come up to you before the water, for in accordance with the principle of the refilling of a vacuum the lighter matter is always the first to succeed to the evacuation.

15. It is not to be wondered at, therefore, that only a very little [nutrient matter] such, namely, as has been accurately elaborated—gets from the stomach into the arteries, since these first become filled with lighter matter. We must understand that there are two kinds of attraction, that by which a vacuum becomes refilled and that caused by appropriateness of quality; air is drawn into bellows in one way, and iron by the lodestone in another. And we must also understand that the traction which results from evacuation acts primarily on what is light, whilst that from appropriateness of quality acts frequently, it may be, on what is heavier (if this should be naturally more nearly related). Therefore, in the case of the heart and the arteries, it is in so far as they are hollow organs, capable of diastole, that they always attract the lighter matter first, while so far as they require nourishment, it is actually into their coats (which are the real bodies of these organs) that the appropriate matter is drawn. Of the blood, then, which is taken into their cavaties when they dilate, that part which is most proper to them and most able to afford nourishment is attracted by their actual coats.

Now, apart from what has been said, the following is sufficient proof that something is taken over from the veins into the arteries. If you will kill an animal by cutting through a number of its large arteries, you will find the veins becoming empty along with the arteries: now, this could never occur if there were not anastomoses between them. Similarly, also, in the heart itself, the thinnest portion of the blood is drawn from the right ventricle into the left, owing to there being perforations in the septum between them: these can be seen for a great part [of their length]; they are like a kind of fossae [pits] with wide mouths, and they get constantly narrower; it is not possible, however, actually to observe their extreme terminations, owing both to the smallness of these and to the fact that when the animal is dead all the parts are chilled and shrunken. Here, too, however, our argument, starting from the principle that nothing is done by Nature in vain, discovers these anastomoses between the ventricles of the heart; for it could not be at random and by chance that there occurred fossae ending thus in narrow terminations.

And secondly [the presence of these anastomoses has been assumed] from the fact that, of the two orifices in the right ventricle, the one conducting blood in and the other out, the former is much the larger. For, the fact that the insertion of the vena cava into the heart is larger than the vein which is inserted into the lungs suggests that not all the blood which the vena cava gives to the heart is driven away again from the heart to the lungs. Nor can it be said that any of the blood is expended in the nourishment of the actual body of the heart, since there is another vein which breaks up in it and which does not take its origin nor get its share of blood from the heart itself. And even if a certain amount is so expended, still the vein leading to the lungs is not to such a slight extent smaller than that inserted into the heart as to make it likely that the blood is used as nutriment for the heart: the disparity is much too great for such an explanation. It is, therefore, clear that something is taken over into the left ventricle.

Moreover, of the two vessels connected with it, that which brings pneuma into it from the lungs is much smaller than the great outgrowing artery from which the arteries all over the body originate; this would suggest that it not merely gets pneuma from the lungs, but that it also gets blood from the right ventricle through the anastomoses mentioned.

WILLIAM HARVEY
(1578–1657)

From *Motion of the Heart and Blood*

Thus nature, ever perfect and divine, doing nothing in vain, has neither given a heart where it was not required, nor produced it before its office had become necessary; but by the same stages in the development of every animal, passing through the constitutions of all, as I may say (ovum, worm, fœtus), it acquires perfection in each. These points will be found elsewhere confirmed by numerous observations on the formation of the fœtus.

Finally, it was not without good grounds that Hippocrates, in his book, "De Corde," intitles it a muscle; as its action is the same, so is its function, viz. to contract and move something else, in this case, the charge of blood.

Farther as in muscles at large, so can we infer the action and use of the heart from the arrangement of its fibres and its general structure. All anatomists admit with Galen that the body of the heart is made up of various courses of fibres running straight, obliquely, and transversely, with reference to one another; but in a heart which has been boiled the arrangement of the fibres is seen to be different: all the fibres in the parietes and septum are circular, as in the sphincters; those, again, which are in the columnæ extend lenthwise, and are oblique longitudinally; and so it comes to pass, that when all the fibres contract simultaneously, the apex of the cone is pulled towards its base by the columnæ, the walls are drawn circularly together into a globe, the whole heart in short is contracted, and the ventricles narrowed; it is therefore impossible not to perceive that, as the action of the organ is so plainly contraction, its function is to propel the blood into the arteries.

Nor are we the less to agree with Aristotle in regard to the sovereignty of the heart; nor are we to inquire whether it receives sense and motion from the brain? whether blood from the liver? whether it be the origin of the veins and of the blood? and more of the same description. They who affirm these propositions against Aristotle, overlook, or do not rightly understand the principal argument, to the effect that the heart is the first part which exists, and that it contains within itself blood, life, sensation, motion, before either the brain or the liver were in being, or had appeared distinctly, or, at all events, before they could perform any function. The heart, ready furnished with its proper organs of motion, like a kind of internal creature, is of a date anterior to the body: first formed, nature willed that it should afterwards fashion, nourish, preserve, complete the entire animal, as its work and dwelling place: the heart, like the prince in a kingdom, in whose hands lie the chief and highest authority, rules over all; it is the original and foundation from which all power is derived, on which all power depends in the animal body.

From *Circulation of the Blood*

How difficult is it to teach those who have no experience, the things of which they have not any knowledge by their senses! And how useless and intractable, and unimpregnable to true science are such auditors! They show the judgment of the blind in regard to colours, of the deaf in reference to concords. Who ever pretended to teach the ebb and flow of the tide, or from a diagram to demonstrate the measurements of the angles and the proportions of the sides of a triangle to a blind man, or to one who had never seen the sea nor a diagram? He who is not conversant with anatomy, inasmuch as he forms no conception of the subject from the evidence of his own eyes, is virtually blind to all that concerns anatomy, and unfit to appreciate what is founded thereon; he knows nothing of that which occupies the attention of the anatomist, nor of the principles inherent in the nature of the things which guide him in his reasonings; facts and inferences as well as their sources are alike unknown to such a one. But no kind of science can possibly flow, save from some pre-existing knowledge of more obvious things; and this is one main reason why our science in regard to the nature of celestial bodies, is so uncertain and conjectural. I would ask of those who profess a knowledge of the causes of all things, why the two eyes keep constantly moving together, up or down, to this side or to that, and not independently, one looking this way, another that; why the two auricles of the heart contract simultaneously, and the like? Are fevers, pestilence, and the wonderful properties of various medicines to be denied because their causes are unknown? Who can tell us why the fœtus in utero, breathing no air up to the tenth month of its existence, is yet not suffocated? Born in the course of the seventh or eighth month, and having once breathed, it is nevertheless speedily suffocated if its respiration be interrupted. Why can the fœtus still contained within the uterus, or enveloped in the membranes, live without respiration; whilst once exposed to the air, unless it breathes it inevitably dies?

Observing that many hesitate to acknowledge the circulation, and others oppose it, because, as I conceive, they have not rightly understood me, I shall here recapitulate briefly what I have said in my work on the Motion of the Heart and Blood. The blood contained in the veins, in its magazine, and where it is collected in largest quantity, viz. in the vena cava, close to the base of the heart and right auricle, gradually increasing in temperature by its internal heat, and becoming attenuated, swells and rises like bodies in a state of fermentation, whereby the auricle being dilated, and then contracting, in virtue of its pulsative power, forthwith delivers its charge into the right ventricle; which being filled, and the systole ensuing, the charge, hindered from returning into the auricle by the tricuspid valves, is forced into the pulmonary artery, which stands open to receive it, and is immediately distended with it. Once in the pulmonary artery, the blood cannot return, by reason of the sigmoid valves; and then the lungs, alternately expanded and contracted during inspiration and expiration, afford it passage by the proper vessels into the pulmonary veins; from the pulmonary veins, the left auricle, acting equally and synchronously with the right auricle, delivers the blood into the left ventricle; which acting harmoniously with the right ventricle, and all regress being prevented by the mitral valves, the blood is projected into the aorta, and consequently impelled into all the arteries of the body. The arteries, filled by this sudden push, as they cannot discharge themselves so speedily, are distended; they receive a shock, or undergo their diastole. But as this process goes on incessantly, I infer that the arteries both of the lungs and of the body at large, under the influence of such a multitude of strokes of the heart and injections of blood, would finally become so over-gorged and distended, that either any further injection must cease, or the vessels would burst, or the whole blood in the body would accumulate within them, were there not an exit provided for it.

The same reasoning is applicable to the ventricles of the heart: distended by the ceaseless action of the auricles, did they not disburthen themselves by the channels of the arteries, they would by and by become over-gorged, and be fixed and made incapable of all motion. Now this, my conclusion, is true and necessary, if my premises be true; but that these are either true or false, our senses must inform us, not our reason—ocular inspection, not any process of the mind.

I maintain further, that the blood in the veins always and everywhere flows from less to greater branches, and from every part towards the heart; whence I gather that the whole charge which the arteries receive, and which is incessantly thrown into them, is delivered to the veins, and flows back by them to the source whence it came. In this way, indeed, is the circulation of the blood established: by an efflux and reflux from and to the heart; the fluid being forcibly projected into the arterial system, and then absorbed and imbibed from every part by the veins, it returns through these in a continuous stream. That all this is so, sense assures us; and necessary inference from the perceptions of sense takes away all occasion for doubt. Lastly, this is what I have striven, by my observations and experiments, to illustrate and make known; I have not endeavoured from causes and probable principles to demonstrate my propositions, but, as of higher authority, to establish them by appeals to sense and experiment, after the manner of anatomists.

And here I would refer to the amount of force, even of violence, which sight and touch make us aware of in the heart and greater arteries; and to the systole and diastole constituting the pulse in the large warm-blooded animals, which I do not say is equal in all the vessels containing blood, nor in all animals that have blood; but which is of such a nature and amount in all, that a flow and rapid passage of the blood through the smaller arteries, the interstices of the tissues, and the branches of the veins, must of necessity take place; and therefore there is a circulation.

JOHN AUBREY
(1626–97)

From *Brief Lives*

WILLIAM HARVEY

William Harvey, Dr. of Physique and Chirurgery, Inventor of the Circulation of the Bloud, was borne at the house which is now the Post-house, a faire stone-built-house, which he gave to Caius college in Cambridge, with some lands there. His brother Eliab would have given any money or exchange for it, because 'twas his father's, and they all borne there; but the Doctor (truly) thought his memory would better be preserved this way, for his brother has left noble seates, and about 3000 pounds per annum, at least.

William Harvey, was always very contemplative, and the first that I heare of that was curious in Anatomie in England. I remember I have heard him say he wrote a booke *De Insectis*, which he had been many yeares about, and had made dissections of Frogges, Toades, and a number of other Animals, and had made curious Observations on them, which papers, together with his goods, in his Lodgings at Whitehall, were plundered at the beginning of the Rebellion, he being for the King, and with him at Oxon; but he often sayd, That of all the losses he sustained, no griefe was so crucifying to him as the losse of these papers, which for love or money he could never retrive or obtaine.

When Charles I by reason of the Tumults left London, he attended him, and was at the fight of Edge-hill with him; and during the fight, the Prince and Duke of Yorke were committed to his care. He told me that he withdrew with them under a hedge, and tooke out of his pockett a booke and read; but he had not read very long before a Bullet of a great Gun grazed on the ground neare him, which made him remove his station.

He told me that Sir Adrian Scrope was dangerously wounded there, and left for dead amongst the dead men, stript; which happened to be the saving of his Life. It was cold, cleer weather, and a frost that night; which staunched his bleeding, and about midnight, or some houres after his hurte, he awaked, and was faine to drawe a dead body upon him for warmeth-sake.

I first sawe him at Oxford, 1642, after Edgehill fight, but was then too young to be acquainted with so great a Doctor. I remember he came severall times to Trinity College to George Bathurst, B.D., who had a Hen to hatch Egges in his chamber, which they dayly opened to discerne the progres and way of Generation. I had not the honour to be acquainted with him till 1651, being my she cosen Montague's physitian and friend. I was at that time bound for Italy (but to my great griefe disswaded by my mother's importunity). He was very communicative, and willing to instruct any that were modest and respectfull to him. And in order to my Journey, gave me, i.e. dictated to me, what to see, what company to keepe, what Bookes to read, how to manage my Studies: in sort, he bid me goe to the Fountain head, and read Aristotle, Cicero, Avicenna, and did call the Neoteriques [modern authors] shitt-breeches.

He wrote a very bad hand, which (with use) I could pretty well read. He understood Greek and Latin pretty well, but was no Critique, and he wrote very bad Latin. The *Circuitis Sanguinis* [Circulation of the Blood] was, as I take it, donne into Latin by Sir George Ent.

At Oxford, he grew acquainted with Dr. Charles Scarborough, then a young Physitian (since by King Charles II Knighted) in whose conversation he much delighted; and whereas before, he marched up and downe with the Army, he tooke him to him and made him ly in his Chamber, and said to him, Prithee leave off thy gunning, and stay here; I will bring thee into practice.

His Majestie King Charles I gave him the Wardenship of Merton Colledge in Oxford, as a reward for his service, but the Times suffered him not to receive or injoy any benefitt by it.

After Oxford was surrendred, which was 24 July 1646, he came to London, and lived with his brother Eliab a rich Merchant in London, who bought, about 1654, Cockainehouse, now (1680) the Excise-office, a noble house, where the Doctor was wont to contemplate on the Leads of the house, and had his severall stations, in regard of the sun, or wind.

He did delight to be in the darke, and told me he could then best contemplate. He had a house heretofore at Combe, in Surrey, a good aire and prospect, where he had Caves made in the Earth, in which in Summer time he delighted to meditate.

Ah! my old Friend Dr. Harvey—I knew him right well. He made me sitt by him 2 or 3 hours together in his meditating apartment discoursing. Why, had he been stiffe, starcht, and retired, as other formall Doctors are, he had known no more than they. From the meanest person, in some way, or other, the learnedst man may learn something. Pride has been one of the greatest stoppers of the Advancement of Learning.

He was far from Bigotry.

He was wont to say that man was but a great, mischievous Baboon.

He had been physitian to the Lord Chancellour Bacon, whom he esteemed much for his witt and style, but would not allow him to be a great Philosopher. Said he to me, *He writes Philosophy like a Lord Chancellor,* speaking in derision; *I have cured him.*

When Doctor Harvey (one of the Physitians College in London) being a Young Man, went to Travel towards Padoa: he went to Dover (with several others) and shewed his Pass, as the rest did, to the Governor there. The Governor told him, that he must not go, but he must keep him Prisoner. The Doctor desired to know for what reason? how he has transgrest. Well it was his Will to have it so. The Pacquet Boat Hoised Sail in the Evening (which was very clear) and the Doctor's Companions in it. There ensued a terrible Storme, and the Pacquet-Boat and all the Passengers were Drown'd: The next day the sad News was brought to Dover. The Doctor was unknown to the Governor, both by Name and Face; but the Night before, the Governor had a perfect Vision in a Dream of Doctor Harvey, who came to pass over to Calais; and that he had a Warning to stop him. This the Governor told the Doctor the next day. The Doctor was a pious good Man, and has several times directed this Story to some of my Acquaintance.

Dr. Harvey told me, and any one if he examines himself will find it to be true, that a man could not fancy—truthfully—that he is imperfect in any part that he has, *verbi gratiâ,* Teeth, Eie, Tongue, Spina dorsi, etc. Natura tends to perfection, and in matters of Generation we ought to consult more with our sense and instinct, then our reason, and prudence, fashion of the country, and Interest. We see what contemptible products are of the prudent politiques; weake, fooles, and ricketty children, scandalls to nature and their country. The Heralds are fooles: *tota errant via* [they are on completely the wrong track]. A blessing goes with a marriage for love upon a strong impulse.

He that marries a widdowe makes himself Cuckold. *Exempli gratia,* if a good Bitch is first warded with a Curre, let her ever after be warded with a dog of good straine and yet she will bring curres as at first, her wombe being first infected with a Curre. So, the children will be like the first Husband (like raysing up children to your brother). So, the Adulterer, though a crime in Law, the children are like the husband.

He would say that we Europeans knew not how to order or governe our Woemen, and that the Turks were the only people used them wisely.

I remember he kept a pretty young wench to wayte on him, which I guesse he made use of for warmeth-sake as King David did, and tooke care of her in his Will, as also of his man servant.

He was very Cholerique; and in his young days wore a dagger (as the fashion then was) but this Dr. would be to apt to draw-out his dagger upon every slight occasion.

I have heard him say, that after his Booke of the *Circulation of the Blood* came-out, that he fell mightily in his Practize, and that 'twas beleeved by the vulgar that he was crack-brained; and all the Physitians were against his Opinion, and envyed him; many wrote against him. With much adoe at last, in about 20 or 30 yeares time, it was received in all the Universities in the world; and, as Mr. Hobbes sayes in his book *De Corpore, he is the only man, perhaps, that ever lived to see his owne Doctrine established in his life-time.*

He was Physitian, and a great Favorite of the Lord High Marshall of England, Thomas Howard Earle of Arundel and Surrey, with whom he travelled as his Physitian in his Ambassade to the Emperor at Vienna. In his Voyage, he would still be making of excursions into the Woods, makeing Observations of strange Trees, and plants, earths, etc., naturalls, and sometimes like to be lost, so that my Lord Ambassador would be really angry with him, for there was not only danger of Thieves, but also of wild beasts.

He was much and often troubled with the Gowte, and his way of Cure was thus; he would then sitt with his Legges bare, if it were a Frost, on the leads of Cockaine-house, putt them into a payle of water, till he was almost dead with cold, and betake himselfe to his Stove, and so 'twas gone.

He was hott-headed, and his thoughts working would many times keepe him from sleepinge; he told me that then his way was to rise out of his Bed and walke about his Chamber in his Shirt till he was pretty coole, i.e. till he began to have a horror, and then returne to bed, and sleepe very comfortably.

He was not tall; but of the lowest stature, round faced, olivaster complexion; little Eie, round, very black, full of spirit; his haire was black as a Raven, but quite white 20 yeares before he dyed.

I remember he was wont to drinke Coffee; which he and his brother Eliab did, before Coffee-houses were in fashion in London.

His practise was not very great towards his later end; he declined it, unlesse to a speciall friend, e.g. my Lady Howland, who had a cancer in her Breast, which he did cutt-off and seared, but at last she dyed of it. He rode on horseback with a Foot-cloath to visitt his Patients, his man following on foote, as the fashion then was, which was very decent, now quite discontinued. (The Judges rode also with their Footecloathes to Westminster-hall, which ended at the death of Sir Robert Hyde, Lord Chief Justice. Anthony Earl of Shafton, would have revived, but severall of the judges being old and ill horsemen would not agree to it.)

All his Profession would allow him to be an excellent Anatomist, but I never heard of any that admired his Therapeutique way. I knew severall practisers in London that would not have given 3d. for one of his Bills; and that a man could hardly tell by one of his Bills what he did aime at. (He did not care for Chymistrey, and was wont to speake against them with an undervalue.)

He had, towards his latter end, a preparation of Opium and I know not what, which he kept in his study to take, if occasion should serve, to putt him out of his paine, and which Sir Charles Scarborough promised to give him; this I beleeve to be true; but doe not at all beleeve that he really did give it him.

Not but that, had he laboured under great Paines, he had been readie enough to have donne it; I doe not deny that it was not according to his Principles upon certain occasions. But the manner of his dyeing was really, and *bonâ fide*, thus, viz. the morning of his death about 10 a clock, he went to speake, and found he had the dead palsey in his Tongue; then he sawe what was to become of him, he knew there was then no hopes of his recovery, so presently sends for his brother and young nephewes to come-up to him, to whom he gives one his Watch ('twas a minute watch with which he made his experiments), to another another thing, etc. as remembrances of him; made a signe to Sambroke, his Apothecary, to lett him blood in the Tongue, which did little or no good; and so ended his dayes. The Palsey did give him an easy Passe-port.

For 20 yeares before he dyed he tooke no manner of care about his worldly concernes, but his brother Eliab, who was a very wise and prudent menager, ordered all not only faithfully, but better then he could have donne himselfe. He dyed worth 20,000 pounds, which he left to his brother Eliab. In his Will he left his old friend Mr. Thomas Hobbes 10 pounds as a token of his Love.

He lies buried in a Vault at Hempsted in Essex, which his brother Eliab Harvey built; he is lapt in lead, and on his brest in great letters

DR. WILLIAM HARVEY.

I was at his Funerall, and helpt to carry him into the Vault.

ABRAHAM COWLEY
(1618–67)

Ode Upon Dr. Harvey

1

Coy Nature, (which remain'd, though aged grown,
  A beauteous virgin still injoy'd by none,
    Nor seen unveil'd by any one)

When *Harvey's* violent passion she did see,
   Began to tremble, and to flee,
Took Sanctuary, like *Daphne*,[1] in a tree:
There *Daphne's* lover stopt, and thought it much
   The very Leaves of her to touch;
But *Harvey*, our *Apollo*, stop't not so,
Into the Bark, and root, he after her did goe:
   No smallest Fibres of a Plant,
For which the eyebeam's Point doth sharpness want,
   His passage after her withstood.
What should she do? through all the moving wood,
Of Lives indow'd with sense, she took her flight;
Harvey pursues, and keeps her still in sight.
But as the Deer long-hunted takes a flood,
She leap't at last into the winding streams of blood;
Of man's *Meander* all the Purple reaches made,
   "Till at the heart she stayd,
   Where turning head, and at a Bay,
Thus, by well-purged ears, was she o're-heard to say.

2

   Here sure I shall be safe (sayd shee)
   None will be able sure to see
   This is my retreat, but only hee,
   Who made both it and mee,
The heart of Man, what Art can e're reveal?
   A Wall Impervious between,
   Divides the very Parts within,
And doth the Heart of man ev'n from it self conceal.
   She spoke, but e're she was aware,
   Harvey was with her there,
And held this slippery *Proteus* in a chain,
"Till all her mighty Mysteries she descry'd;
Which from his wit th' attempt before to hide,
Was the first Thing that Nature did in vain.

---

[1] The chaste goddess who transformed herself to escape love.

3

He the young Practise of New life did see,
Whil'st to conceal its toylsome Poverty,
It for a Living wrought, both hard, and privately.
  Before the Liver understood
  The noble Scarlet Dye of Blood,
  Before one drop was by it made,
Or brought into it, to set up the Trade;
Before the untaught Heart began to beat
The tuneful March to vital Heat,
From all the Souls that living Buildings rear,
Whether implyd for earth, or sea, or air,
Whether it in the womb or egg be wrought,
A strict account to him is hourly brought,
  How the Great Fabrick do's proceed;
What time and what materials it do's need.
He so exactly do's the work survey,
As if he hir'd the workers by the day.

4

Thus *Harvey* sought for truth in truth's own Book,
The creatures, which by God himself was writ;
    And wisely thought 'twas fit,
Not to read Comments only upon it,
But on th' original it self to look.
Methinks in Art's great Circle others stand
  Lockt up together, Hand in Hand;
  Ev'ry one leads as he is led;
    The same bare path they tread,
And Dance, like Fairies, a fantastick round,
But neither change their motion, nor their ground:
Had *Harvey* to this Road confin'd his wit,
His noble Circle of the Blood, had been untrodden yet.
Great Doctor! Th' art of Curing's cur'd by thee,
  We now thy Patient Physick see,
From all inveterate diseases free;
  Purg'd of old errors by thy Care,

New-dieted, put forth to clearer ayr.
   It now will strong, and healthful prove;
   It self before Lethargick lay and could not move.

5

These Vseful secrets to his Pen we owe,
And thousands more 'twas ready to bestow;
Of which, a Barbarous War's unlearned Rage
   Has robb'd the Ruin'd Age;
O cruel loss ! as if the Golden Fleece,
   With so much cost, and labour bought,
And from afarr by a Great *Hero*[2] Brought,
   Had sunk even in the Ports of *Greece*.
O cursed warre ! who can forgive thee this?
   Houses and towns may rise again,
   And ten times easier it is
To re-build *Pauls*,[3] than any work of his,
That mighty task none but himself can doe,
   Nay, scarce himself too now;
For though his Wit the force of Age withstand,
His Body, alas! and time it must command,
And Nature now, so long by him surpass't,
Will sure have her Revenge on him at last.

[2] Jason.
[3] St. Paul's cathedral, London, burned in the great fire of 1666 and was rebuilt by Christopher Wren.

HENRY BAKER

(1698–1774)

From *Of Microscopes and the Discoveries Made Thereby*

Upon viewing the *milt* or *semen Masculinum* of a living Cod-fish with a *Microscope*, such Numbers of *Animalcules* with long Tails were found therein, that at least ten thousand of them were supposed to exist in the Quantity of a Grain of Sand. Whence Mr. Leeuwenhoek argues, that the Milt of that single Cod-fish contained more living *Animalcules* than there are People alive upon the Face of the whole Earth at one and the same Time: for he computes that one hundred Grains of Sand make the Diameter of an Inch; wherefore in a cubic Inch there will be a million of such Sands. And as he found the Milt of the Cod-fish to be about fifteen cubic Inches, it must contain fifteen millions of Quantities as big as a Grain of Sand. Now if each of these Quantities contain ten thousand *Animalcules*, there must be in the Whole one hundred and fifty thousand millions.

Then, to find out, in a probable Manner, the Number of People living upon the whole Earth at one Time; he reckons, that in a great Circle there are five thousand four hundred *Dutch* square Miles; whence he calculates the Surface of the Earth to contain nine millions two hundred seventy-six thousand two hundred and eighteen such square Miles: and supposing one third of the Whole, or three millions ninety-two thousand and seventy-two Miles, to be dry Land; and of this, two thirds, or two millions sixty-one thousand three hundred and eighty-two Miles, to be inhabited: and supposing farther, that *Holland* and *West-Friesland* are twenty-two Miles long and seven broad, which make one hundred and fifty-four square Miles; the habitable Part of the World is thirteen thousand three hundred and eighty-five Times the Bigness of *Holland* and *West-Friesland*.

Now, if the People in these two Provinces be supposed a million, and if all the other inhabited Parts of the World were as populous as these (which is highly improbable), there would be thirteen thousand three hundred and eighty-five millions of People on the Face of the whole Earth: but the Milt of this Cod-fish contained one hundred and fifty thousand millions of *Animalcules*, which is ten Times more than the Number of all Mankind.

*   *   *

The human *Semen* has likewise been viewed by the *Microscope*, and found no less plentifully stocked with life than that of other animals: for more than ten thousand living Creatures were seen, by Mr. Leeuwenhoek, moving in no larger a Quantity of the fluid Part thereof than the Bigness of a Grain of Sand; and in the thicker Part they were so thronged together, that they could not move for one another. Their Size was smaller than the red Globules of the Blood, and even less than the millionth Part of a Grain of Sand. The Bodies of them are roundish, somewhat flat before, but ending sharp behind, with Tails exceedingly transparent, five or six Times longer, and about five Times more slender, than their Bodies. They move themselves along by the violent Agitation of their Tails, in various Bendings, after the Manner that Eels or Serpents swim: and sometimes their Tails are moved thus eight or ten Times in getting forwards the Diameter of a Hair.

It is wonderful to consider the Minuteness of these little Animals, and particularly the amazing Slenderness of their Tails: which must, notwithstanding, be furnished with as many Joints as the Tails of larger Creatures, since they are able to move them with great Agility: and, besides, every one of these Joints must be provided with its proper Muscles, Nerves, Arteries, and Veins; and also with Fluids circulating thro' them, and supplying them with Nourishment, Strength, and Motion. In short, the Mind loses itself in contemplating a Minuteness beyond all human Conception; tho' Reason tells us it certainly must be. I remember Dr. Power[4] has a fine Passage to this Purpose in the Preface of his experiments: "It has often seemed to me (says he) an ordinary Probability, and something more than Fancy (how paradoxical soever the Conjecture may seem), to think that the least Bodies we are able to see with our naked Eyes are but *middle Proportionals*, as it were, betwixt the *greatest* and the *smallest* Bodies in Nature; which two Extremes lie equally beyond the Reach of human Sensation.—For, as on one Side they are but narrow Souls, and not worthy the Name of *Philosophers*, that think any Body can be *too great* or *too vast* in its Dimensions: so likewise are they as inapprehensive, and of the same Litter with the former, that, on the other Side, think the Particles of Matter may be *too little*, or that Nature is stinted at an Atom, and must have a *Non ultra* of her Subdivisions."

EMILY DICKINSON

(1830–86)

"Faith" Is a Fine Invention

"Faith" is a fine invention
When Gentlemen can *see*—
But *Microscopes* are prudent
In an Emergency.

---

[4]Henry Power (1623–68) did experiments and wrote a short poem on the microscope, four lines from which follow:

> For what a better, fitter, gift Could bee
> in this world's Aged Luciosity,
> To Help our Blindnesse so as to devize
> a paire of new & Artificial eyes.

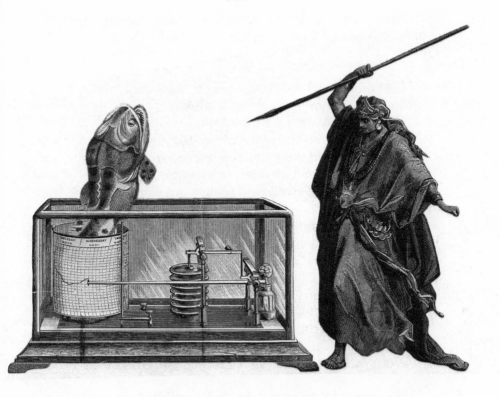

IZAAK WALTON

(1593–1683)

From *The Compleat Angler*

And the great naturalist Pliny says, "That nature's great and wonderful power is more demonstrated in the sea than on the land." And this may appear by the numerous and various creatures inhabiting both in and about that element; as to the reader of Gesner,[5] Rondeletius,[6] Pliny, Ausonius,[7] Aristotle,

[5]Konrad Gesner (1516–65), a writer of natural history compendia.
[6]Guillaume Rondelet (1507–65), the author of an encyclopedia of aquatic animals.
[7]Decimus Magnus Ausonius (A.D. 310–394?).

and others, may be demonstrated. But I will sweeten this discourse also out of a contemplation in divine Du Bartas,[8] who says:

> God quickened in the sea, and in the rivers
> So many fishes of so many features
> That in the waters we may see all creatures.
> Even all that on the earth are to be found.
> As if the world were in deep waters drowned.
> For seas—as well as skies—have sun, moon, stars,
> As well as air—swallows, rooks, and stares;
> As well as earth—wines, roses, nettles, melons,
> Mushrooms, pinks, gilliflowers, and many millions
> Of other plants more rare, more strange than these,
> As very fishes living in the seas;
> As also rams, calves, horses, hares and hogs,
> Wolves, urchins, lions, elephants, and dogs;
> Yea, men and maids, and which I most admire,
> The mitred bishop and the cowled friar.
> Of which, examples, but a few years since,
> Were shown the Norway and Polonian prince.

These seem to be wonders, but have had so many confirmations from men of learning and credit that you need not doubt them. Nor are the number nor the various shapes of fishes more strange or more fit for contemplation than their different natures, inclinations, and actions; concerning which I shall beg your patient ear a little longer.

The cuttle-fish will cast a long gut out of her throat, which, like as an angler doth his line, she sendeth forth and pulleth in again at her pleasure according as she sees some little fish come near to her; and the cuttle-fish, being then hid in the gravel, lets the smaller fish nibble and bite the end of it, at which time she by little and little draws the smaller fish so near to her that she may leap upon her, and then catches and devours her. And for this reason some have called this fish the sea-angler.

[8]Guillaume de Salluste (1544–90) Seigneur Du Bartas.

And there is a fish called a hermit [crab], that at a certain age gets into a dead fish's shell and like a hermit dwells there alone, studying the wind and weather, and so turns her shell that she makes it defend her from the injuries that they would bring upon her.

There is also a fish called by Elian (in his 9th book of living creatures, chap. 16) the Adonis, or darling of the sea, so called because it is a loving and innocent fish, a fish that hurts nothing that hath life and is at peace with all the numerous inhabitants of that vast watery element. And truly I think most anglers are so disposed to most of mankind.

And there are also lustful and chaste fishes; of which I shall give you examples.

And first, what Du Bartas says of a fish called the sargus; which—because none can express it better than he does—I shall give you in his own words, supposing it shall not have the less credit for being verse, for he hath gathered this and other observations out of authors that have been great and industrious searchers into the secrets of nature:

> The adult'rous sargus doth not only change
> Wives every day in the deep streams, but, strange,
> As if the honey of sea-love delight
> Could not suffice his ranging appetite,
> Goes courting she-goats on the grassy shore,
> Horning their husbands that had horns before.

And the same author writes concerning the cantharus that which you shall also hear in his own words:

> But, contrary, the constant cantharus
> Is ever constant to his faithful spouse,
> In nuptial duties spending his chaste life,
> Never loves any but his own dear wife.

JOSEPH ADDISON

(1672–1719)

From *The Spectator*

NO. 519

Inde hominum pecudumque genus, vitaeque volantum,
Et quae marmoreo fert monstra sub aequore pontus.—Virgil
[Hence man and beasts the breath of life obtain,
And birds of air, and monster of the main.]—Dryden

Though there is a great deal of Pleasure in contemplating the
Material World, by which I mean that System of Bodies into
which Nature has so curiously wrought the Mass of dead Matter, with the several Relations which those bodies bear to one
another; there is still, methinks, something more wonderful
and surprizing in Contemplations on the World of Life, by
which I mean all those Animals with which every Part of the
Universe is furnished. The Material World is only the Shell of
the Universe: The World of Life are its Inhabitants.

If we consider those Parts of the Material World which lie
the nearest to us, and are therefore subject to our Observations and Enquiries, it is amazing to consider the Infinity of
Animals with which it is stocked. Every part of Matter is peo-
pled: Every green Leaf swarms with Inhabitants. There is
scarce a single Humour in the Body of a Man, or of any other
Animal, in which our Glasses do not discover Myriads of living Creatures. The Surface of Animals is also covered with

other Animals, which are in the same manner the Basis of other Animals that live upon it; nay, we find in the most solid Bodies, as in Marble it self, innumerable Cells and Cavities that are crouded with such imperceptible Inhabitants, as are too little for the naked Eye to discover. On the other hand, if we look into the more bulky Parts of Nature, we see the Seas, Lakes and Rivers teeming with numberless Kinds of living Creatures: We find every Mountain and Marsh, Wilderness and Wood, plentifully stocked with Birds and Beasts, and every part of Matter affording proper Necessaries and Conveniencies for the Livelihood of Multitudes which inhabit it.

The Author of the *Plurality of Worlds*[9] draws a very good Argument from this Consideration, for the *peopling* of every Planet, as indeed it seems very probable from the Analogy of Reason, that if no part of Matter, which we are acquainted with, lies waste and useless, those great Bodies which are at such a Distance from us should not be desart and unpeopled, but rather that they should be furnished with Beings adapted to their respective Situations.

Existence is a Blessing to those Beings only which are endowed with Perception, and is, in a manner, thrown away upon dead Matter, any further than as it is subservient to Beings which are conscious of their Existence. Accordingly we find, from the Bodies which lie under our Observation, that Matter is only made as the Basis and Support of Animals, and that there is no more of the one, than what is necessary for the Existence of the other.

Infinite Goodness is of so communicative a Nature, that it seems to delight in the conferring of Existence upon every degree of Perceptive Being. As this is a Speculation which I have often pursued with great Pleasure to my self, I shall enlarge farther upon it, by considering that part of the Scale of Beings which comes within our Knowledge.

There are some living Creatures which are raised but just above dead Matter. To mention only that Species of Shellfish, which are formed in the Fashion of a Cone, that grow to the Surface of several Rocks, and immediately die upon their

[9]Bernard Le Bovier de Fontenelle (1657–1757).

being severed from the Place where they grow. There are many other Creatures but one Remove from these, which have no other Sense besides that of Feeling and Taste. Others have still an additional one of Hearing; others of Smell, and others of Sight. It is wonderful to observe, by what a gradual Progress the World of Life advances through a prodigious Variety of Species, before a Creature is formed that is compleat in all its Senses, and even among these there is such a different degree of Perfection in the Sense, which one Animal enjoys beyond what appears in another, that though the Sense in different Animals be distinguished by the same common Denomination, it seems almost of a different Nature. If after this we look into the several inward Perfections of Cunning and Sagacity, or what we generally call Instinct, we find them rising after the same manner, imperceptibly one above another, and receiving additional Improvements, according to the Species in which they are implanted. This Progress in Nature is so very gradual, that the most perfect of an inferior Species comes very near to the most imperfect of that which is immediately above it.

The exuberant and overflowing Goodness of the Supream Being, whose Mercy extends to all his Works, is plainly seen, as I have before hinted, from his having made so very little Matter, at least what falls within our Knowledge, that does not Swarm with Life: Nor is his Goodness less seen in the Diversity than in the Multitude of living Creatures. Had he only made one Species of Animals, none of the rest would have enjoyed the Happiness of Existence; he has, therefore, *specified* in his Creation every degree of Life, every Capacity of Being. The whole Chasm in Nature, from a Plant to a Man, is filled up with diverse Kinds of Creatures, rising one over another, by such a gentle and easie Ascent, that the little Transitions and Deviations from one Species to another, are almost insensible. This intermediate Space is so well husbanded and managed, that there is scarce a degree of Perception which does not appear in some one part of the World of Life. Is the Goodness or Wisdom of the Divine Being more manifested in this his Proceeding?

There is a Consequence, besides those I have already men-

tioned, which seems very naturally deducible from the foregoing Considerations. If the Scale of Being rises by such a
regular Progress, so high as Man, we may by a Parity of Reason suppose that it still proceeds gradually through those
Beings which are of a Superior Nature to him, since there is
an infinitely greater Space and Room for different Degrees of
Perfection, between the Supreme Being and Man, than between Man and the most despicable Insect. This Consequence
of so great a Variety of Beings which are superior to us, from
that Variety which is inferior to us, is made by Mr. *Lock*, in a
Passage which I shall here set down, after having premised,
that notwithstanding there is such infinite room between Man
and his Maker for the Creative Power to exert it self in, it is
impossible that it should ever be filled up, since there will be
still an infinite Gap or Distance between the highest created
Being, and the Power which produced him.

That there should be more Species *of intelligent Creatures above
us, than there are of sensible and material below us, is probable to me
from hence; That in all the visible corporeal World, we see no
Chasms, or no Gaps. All quite down from us, the descent is by easie
steps, and a continued series of things, that in each remove, differ
very little one from the other. There are Fishes that have Wings, and
are not Strangers to the airy Region; and there are some Birds, that
are Inhabitants of the Water; whose Blood is cold as Fishes, and their
Flesh so like in Taste, that the scrupulous are allowed them on Fish-
days. There are Animals so near of kin both to birds and Beasts, that
they are in the middle between both: Amphibious Animals link the
Terrestrial and Aquatique together; Seals live at Land and at Sea,
and Porpoises have the warm Blood and Entrails of a Hog, not to
mention what is confidently reported of Mermaids or Sea-men. There
are some Brutes, that seem to have as much Knowledge and Reason,
as some that are called Men; and the Animal and Vegetable Kingdoms are so nearly joyn'd, that if you will take the lowest of one, and
the highest of the other, there will scarce be perceived any great difference between them; and so on till we come to the lowest and the
most inorganical parts of Matter, we shall find every where that the
several* Species *are linked together, and differ but in almost insensible degrees. And when we consider the infinite Power and Wisdom of
the Maker, we have reason to think, that it is suitable to the magnifi-*

cent Harmony of the Universe, and the great Design and infinite Goodness of the Architect, that the Species of Creatures should also, by gentle degrees, Ascend upward from us toward his infinite Perfection, as we see they gradually descend from us downward: Which if it be probable, we have reason then to be persuaded, that there are far more Species of Creatures above us, than there are beneath; we being in degrees of perfection much more remote from the infinite Being of God, than we are from the lowest state of Being, and that which approaches nearest to nothing. And yet of all those distinct Species, we have no clear distinct Ideas [John Locke].

In this System of Being, there is no Creature so wonderful in its Nature, and which so much deserves our particular Attention, as Man, who fills up the middle Space between the Animal and Intellectual Nature, the visible and invisible World, and is that Link in the Chain of Beings which has been often termed the *nexus utriusque mundi.* So that he, who in one Respect is associated with Angels and Arch-Angels, may look upon a Being of infinite Perfection as his Father, and the highest Order of Spirits as his Brethren, and may in another Respect say to *Corruption, thou art my Father, and to the Worm, thou art my Mother and my Sister.*

ALFRED NOYES
(1880–1958)

From *The Torch-Bearers*

Beware of old Linnaeus,
   The Man of the Linden-tree,
So beautiful, bright and early
   He brushed away the dews
He found the wicked wild-flowers
   All courting there in twos.

ERASMUS DARWIN
(1731–1802)

From *Botanic Garden*

PART II
LOVES OF THE PLANTS

Canto II

When the young Hours, amid her tangled hair,
Wove the fresh rose-bud, and the lily fair,
Proud Gloriosa[10] led *three* chosen swains,
the blushing captives of her virgin chains—
When Time's rude hand a bark of wrinkles spread
Round her weak limbs, and silver'd o'er her head,
Three other youths her riper years engage,
The flatter'd victims of her wily age.

So, in her wane of beauty, Ninon won
With fatal smiles her gay unconscious son.—
Clasp'd in his arms, she own'd a mother's name,—
"Desist, rash youth! restrain your impious flame,
"First on that bed your infant-form was press'd,
"Born by my throes, and nurtured at my breast."—
Back as from death he sprung, with wild amaze
Fierce on the fair he fix'd his ardent gaze;
Dropp'd on one knee, his frantic arms outspread,
And stole a guilty glance toward the bed;
Then breathed from quivering lips a whisper'd vow,
And bent on heaven his pale repentant brow;

---

[10]Gloriosa. Superba (*summer glory lily*), six males (*stamens*), one female (*pistil*).

"Thus, thus!" he cried, and plunged the furious dart,
And life and love gush'd, mingled, from his heart.

*   *   *

Canto II

Papyra,[11] throned upon the banks of Nile,
Spread her smooth leaf, and waved her silver style.
—The storied pyramid, the laurel'd bust,
The trophy'd arch had crumbled into dust;
The sacred symbol, and the epic song
(Unknown the character, forgot the tongue,)
With each unconquer'd chief, or sainted maid,
Sunk undistinguish'd in Oblivion's shade.
Sad o'er the scatter'd ruins Genius sigh'd,
And infant Arts but learn'd to lisp, and died.
Till to astonish'd realms Papyra taught
To paint in mystic colours Sound and Thought,
With Wisdom's voice to print the page sublime,
And mark in adamant the steps of Time.
—*Three* favour'd youths her soft attention share,
The fond disciples of the studious Fair,
Hear her sweet voice, the golden process prove;
Gaze, as they learn; and as they listen, love.
The first from Alpha to Omega joins
The letter'd tribes along the level lines;
Weighs with nice ear the vowel, liquid, surd,[12]
And breaks in syllables the volant word.
Then forms *the next* upon the marshall'd plain,
In deepening ranks, his dexterous cypher-train;
And counts, as wheel the decimating bands,
The dews of Egypt, or Arabia's sands.
And then *the third,* on four concordant lines,
Prints the lone crotchet, and the quaver joins;
Marks the gay trill, the solemn pause inscribes,
And parts with bars the undulating tribes.
Pleased, round her cane-wove throne, the applauding crowd

[11]*Cyperus Papyrus.* Three males, one female.
[12]In phonetics a voiceless breath consonant. The three youths represented court
Papyra for writing, arithmetic and music notation.

Clapp'd their rude hands, their swarthy foreheads bow'd;
With loud acclaim, "a present God!" they cry'd,
"A present God!" rebellowing shores reply'd.—
Then peal'd at intervals, with mingled swell,
The echoing harp, shrill clarion, horn, and shell;
While Bards ecstatic, bending o'er the lyre,
Struck deeper chords, and wing'd the song with fire.
Then mark'd Astronomers, with keener eyes,
The Moon's refulgent journey through the skies;
Watch'd the swift Comets urge their blazing cars,
And weigh'd the Sun with his revolving Stars.
High raised the Chemists their Hermetic wands,
(And changing forms obey'd their waving hands,)
Her treasured Gold from Earth's deep chambers tore,
Or fused and harden'd her chalybeate ore.
All with bent knee from fair Papyra claim,
Wove by her hands, the wreath of deathless fame.
—Exulting Genius crown'd his darling child,
The young Arts clasp'd her knees, and Virtue smiled.

* * *

Canto IV

Warm with sweet blushes bright Galantha[13] glows,
And prints with frolic step the melting snows;
O'er silent floods, white hills, and glittering meads
Six rival swains the playful beauty leads,
Chides with her dulcet voice the tardy Spring,
Bids slumbering Zephyr stretch his folded wing,
Wakes the hoarse Cuckoo in his gloomy cave,
And calls the wondering Dormouse from his grave,
Bids the mute Redbreast cheer the budding grove,
And plaintive Ringdove tune her notes to love.

[13]Galanthus. Nivalis *(snow drop)*. Six males, one female.

─── THE REVEREND GILBERT WHITE ───
(1720–93)

From *The Natural History of Selborne*

LETTER III

The fossil shells of this district, and sorts of stone, such as have fallen within my observation, must not be passed over in silence. And first I must mention, as a great curiosity, a specimen that was ploughed up in the chalky fields, near the side of the down, and given to me for the singularity of its appearance, which, to an incurious eye, seems like a petrified fish of about four inches long, the *cardo* passing for a head and mouth. It is in reality a bivalve of the Linnæan genus of *Mytilus*, and the species of Crista Galli; called by Lister,[14]

[14]Martin Lister (1639–1712), pioneer in the taxonomy of mollusks.

Rastellum; by Rumphius,[15] *Ostreum plicatum minus;* by D'Argenville,[16] *Auris Porci,* s. *Crista Galli,* and by those who make collections *cock's comb.* Though I applied to several such in London, I could never meet with an entire specimen; nor could I ever find in books any engraving from a perfect one. In the superb museum at Leicester-house, permission was given me to examine for this article; and though I was disappointed as to the fossil, I was highly gratified with the sight of several of the shells themselves in high preservation. This bivalve is only known to inhabit the Indian Ocean, where it fixes itself to a zoophyte, known by the name *Gorgonia.* The curious foldings of the suture, the one into the other, the alternate flutings or grooves, and the curved form of my specimen being much easier expressed by the pencil than by words, I have caused it to be drawn and engraved.

Cornua Ammonis[17] are very common about this village. As we were cutting an inclining path up the Hanger,[18] the labourers found them frequently on that steep, just under the soil, in the chalk, and of considerable size. In the lane above Well-head, in the way to Emshot, they abound in the bank, in a darkish sort of marl; and are usually very small and soft: but in Clay's Pond, a little farther on, at the end of the pit, where the soil is dug out for manure, I have occasionally observed them of large dimensions, perhaps fourteen or sixteen inches in diameter. But as these did not consist of firm stone, but were formed of a kind of *terra lapidosa,* or hardened clay, as soon as they were exposed to the rains and frost they mouldered away. These seemed as if they were a very recent production. In the chalk-pit, at the north-west end of the Hanger, large *nautili* are sometimes observed.

In the very thickest strata of our freestone, and at considerable depths, well-diggers often find large scallops or *pectines,* having both shells deeply striated, and ridged and furrowed alternately. They are highly impregnated with, if not wholly composed of, the stone of the quarry.

[15]George Eberhard Rumpf (1627–1702), Dutch naturalist who worked in the East Indies.
[16]Antoine-Joseph Dezallier D'Argenville (1680–1765), collector and engraver of natural history.
[17]Ammonites.
[18]A wooded walk on the hill overlooking White's house.

Letter XXIV

The *scarabœus fullo* [beetle] I know very well, having seen it in collections; but have never been able to discover one wild in its natural state. Mr. Banks[19] told me he thought it might be found on the sea-coast.

On the thirteenth of April I went to the sheep-down, where the ring-ousels have been observed to make their appearance at spring and fall, in their way perhaps to the north or south; and was much pleased to see three birds about the usual spot. We shot a cock and a hen; they were plump and in high condition. The hen had but very small rudiments of eggs within her, which proves they are late breeders; whereas those species of the thrush kind that remain with us the whole year have fledged young before that time. In their crops was nothing very distinguishable, but somewhat that seemed like blades of vegetables nearly digested. In autumn they feed on haws and yew-berries, and in the spring on ivy-berries. I dressed one of these birds, and found it juicy and well-flavoured. It is remarkable that they make but a few days' stay in their spring visit, but rest near a fortnight at Michaelmas. These birds, from the observations of three springs and two autumns, are most punctual in their return; and exhibit a new migration unnoticed by the writers, who supposed they never were to be seen in any of the southern counties.

One of my neighbours lately brought me a new *salicaria* [sedge warbler] which at first I suspected might have proved your willow-lark but, on a nicer examination, it answered much better to the description of that species which you shot at Revesby, in Lincolnshire. My bird I describe thus: "It is a size less than the grasshopper-lark; the head, back, and coverts of the wings of a dusky brown, without those dark spots of the grasshopper-lark; over each eye is a milk-white stroke; the chin and throat are white, and the under parts of a yellowish white; the rump is tawny and the feathers of the tail sharp-pointed; the bill is dusky and sharp, and the legs are

---

[19]Sir Joseph Banks (1743–1820), the naturalist who accompanied Captain Cook on his voyages.

dusky; the hinder claw long and crooked." The person that shot it says that it sung so like a reed-sparrow that he took it for one; and that it sings all night; but this account merits further inquiry. For my part, I suspect it is a second sort of *locustella*, hinted at by Dr. Derham in Ray's[20] *Letters*. . . . He also procured me a grasshopper-lark.

The question that you put with regard to those genera of animals that are peculiar to America, viz. how they came there, and whence? is too puzzling for me to answer; and yet so obvious as often to have struck me with wonder. If one looks into the writers on that subject little satisfaction is to be found. Ingenious men will readily advance plausible arguments to support whatever theory they shall choose to maintain; but then the misfortune is, every one's hypothesis is each as good as another's, since they are all founded on conjecture. The late writers of this sort, in whom may be seen all the arguments of those that have gone before, as I remember, stock America from the western coast of Africa and south of Europe; and then break down the Isthmus that bridged over the Atlantic. But this is making use of a violent piece of machinery: it is a difficulty worthy of the interposition of a god! *"Incredulus odi."*

The Naturalist's Summer-Evening Walk

. . . equidem credo, quia sit divinitus illis Ingenium.
Virgil Georgics

When day declining sheds a milder gleam,
What time the may-fly haunts the pool or stream;
When the still owl skims round the grassy mead,
What time the timorous hare limps forth to feed;
Then be the time to steal adown the vale,
And listen to the vagrant cuckoo's tale;
To hear the clamorous curlew call his mate,
Or the soft quail his tender pain relate;
To see the swallow sweep the dark'ning plain

[20]John Ray (1627–1705), naturalist, a major authority of White's age.

Belated, to support her infant train;
To mark the swift in rapid giddy ring
Dash round the steeple, unsubdu'd of wing:
Amusive birds!—say where your hid retreat
When the frost rages and the tempests beat;
Whence your return, by such nice instinct led,
When spring, soft season, lifts her bloomy head?
Such baffled searches mock man's prying pride,
The GOD of NATURE is your secret guide!

While deep'ning shades obscure the face of day
To yonder bench, leaf-shelter'd, let us stray,
Till blended objects fail the swimming sight,
And all the fading landscape sinks in night;
To hear the drowsy dor come brushing by
With buzzing wing, or the shrill cricket cry;
To see the feeding bat glance through the wood;
To catch the distant falling of the flood;
While o'er the cliff th' awakened churn-owl hung
Through the still gloom protracts his chattering song;
While high in air, and pois'd upon his wings,
Unseen, the soft enamour'd woodlark sings:
These NATURE's works, the curious mind employ,
    Inspire a soothing melancholy joy:
As fancy warms, a pleasing kind of pain
Steals o'er the cheek, and thrills the creeping vein!
    Each rural sight, each sound, each smell combine;
The tinkling sheep-bell, or the breath of kine;
The new-mown hay that scents the swelling breeze,
Or cottage-chimney smoking through the trees.
    The chilling night-dews fall: away, retire;
For see, the glow-worm lights her amorous fire!
Thus, ere night's veil had half obscured the sky,
Th' impatient damsel hung her lamp on high:
True to signal, by love's meteor led,
Leander hasten'd to his Hero's bed.

SIR CHARLES LYELL

(1797–1875)

From *Principles of Geology*

One consequence of undervaluing greatly the quantity of past time, is the apparent coincidence which it occasions of events necessarily disconnected, or which are so unusual, that it would be inconsistent with all calculation of chances to suppose them to happen at one and the same time. When the unlooked-for association of such rare phenomena is witnessed in the present course of nature, it scarcely ever fails to excite a suspicion of the preternatural in those minds which are not firmly convinced of the uniform agency of secondary causes;—as if the death of some individual in whose fate they are interested happens to be accompanied by the appearance of a luminous meteor, or a comet, or the shock of an earthquake. It would be only necessary to multiply such coincidences indefinitely, and the mind of every philosopher would be disturbed. Now it would be difficult to exaggerate the number of physical events, many of them most rare and unconnected in their nature, which were imagined by the Woodwardian hypothesis[21] to have happened in the course of a few months: and numerous other examples might be found of popular geological theories, which require us to imagine that a long succession of events happened in a brief and almost momentary period.

Another liability to error, very nearly allied to the former, arises from the frequent contact of geological monuments referring to very distant periods of time. We often behold, at one glance, the effects of causes which have acted at times incalculably remote, and yet there may be no striking circumstances to mark the occurrence of a great chasm in the chronological series of Nature's archives. In the vast interval of time which may really have elapsed between the results of operations thus compared, the physical condition of the earth may, by slow and insensible modifications, have become entirely altered; one or more races of organic beings may have passed away, and yet have left behind, in the particular region under contemplation, no trace of their existence.

* * *

The general inference drawn from the study and comparison of the various groups [of sedimentary strata], arranged in chronological order, is this: that at successive periods distinct tribes of animals and plants have inhabited

[21]John Woodward (1665?–1728), author of An Essay, Toward a Natural History of the Earth, which dates fossils from Noah's flood.

the land and waters, and that the organic types of the newer formations are more analogous to species now existing than those of more ancient rocks. If we then turn to the present state of the animate creation, and enquire whether it has now become fixed and stationary, we discover that, on the contrary, it is in a state of continual flux—that there are many causes in action which tend to the extinction of species, and which are conclusive against the doctrine of their unlimited durability.

There are also causes which give rise to new varieties and races in plants and animals, and new forms are continually supplanting others which had endured for ages. But natural history has been successfully cultivated for so short a period, that a few examples only of local, and perhaps but one or two of absolute, extirpation of species can as yet be proved, and these only where the interference of man has been conspicuous. It will nevertheless appear evident, from the facts and arguments detailed in the chapters which treat of the geographical distribution of species . . . that man is not the only exterminating agent; and that, independently of his intervention, the annihilation of species is promoted by the multiplication and gradual diffusion of every animal or plant. It will also appear that every alteration in the physical geography and climate of the globe cannot fail to have the same tendency. If we proceed still farther, and enquire whether new species are substituted from time to time for those which die out, we find that the successive introduction of new forms appears to have been a constant part of the economy of the terrestrial system, and if we have no direct proof of the fact it is because the changes take place so slowly as not to come within the period of exact scientific observation. To enable the reader to appreciate the gradual manner in which a passage may have taken place from an extinct fauna to that now living, I shall say a few words on the fossils of successive Tertiary periods. When we trace the series of formations from the more ancient to the more modern, it is in these Tertiary deposits that we first meet with assemblages of organic remains having a near analogy to the fauna of certain parts of the globe in our own time. In the Eocene, or oldest subdivisions, some few of the testacea [single-celled rhizopods with shells] belong to existing species although almost all of them, and apparently all the associated vertebrata, are now extinct. These Eocene strata are succeeded by a great number of more modern deposits, which depart gradually in the character of their fossils from the Eocene type, and approach more and more to that of the living creation. In the present state of science, it is chiefly by the aid of shells that we are enabled to arrive at these results, for of all classes the testacea are the most generally diffused in a fossil state, and may be called the medals principally employed by nature in recording the chronology of past events. In the Upper Miocene rocks . . . we begin to find considerable number, although still a minority, of recent species, intermixed with some fossils common to the preceding, or Eocene, epoch. We then arrive at the Pliocene strata, in which species now contemporary with man begin to preponderate, and in the newest of which nine-tenths of the fossils agree with species still inhabiting the neighbouring sea. It is in the Post-Tertiary strata, where all the shells agree with species now living, that we have discovered the first or earliest known remnants of man associated with the bones of quadrupeds, some of which are of extinct species.

*   *   *

It may undoubtedly be said that strata have been always forming somewhere, and therefore at every moment of past time Nature has added a page to her archives.

——— ALFRED RUSSEL WALLACE ———
(1823–1913)

From *On the Law Which Has Regulated the Introduction of New Species* (1855)

GEOGRAPHICAL DISTRIBUTION DEPENDENT ON GEOLOGIC CHANGES

Every naturalist who has directed his attention to the subject of the geographical distribution of animals and plants, must have been interested in the singular facts which it presents. Many of these facts are quite different from what would have been anticipated, and have hitherto been considered as highly curious, but quite inexplicable. None of the explanations attempted from the time of Linnæus are now considered at all satisfactory; none of them have given a cause sufficient to account for the facts known at the time, or comprehensive enough to include all the new facts which have since been, and are daily being added. Of late years, however, a great light has been thrown upon the subject by geological investigations, which have shown that the present state of the earth and of the organisms now inhabiting it, is but the last stage of a long and uninterrupted series of changes which it has undergone, and consequently, that to endeavour to explain and account for its present condition without any reference to those changes (as has frequently been done) must lead to very imperfect and erroneous conclusions.

The facts proved by geology are briefly these:—That during an immense, but unknown period, the surface of the earth has undergone successive changes; land has sunk beneath the ocean, while fresh land has risen up from it; mountain chains have been elevated; islands have been formed into continents, and continents submerged till they have become islands; and these changes have taken place, not once merely, but perhaps hundreds, perhaps thousands of times:—That all these operations have been more or less continuous, but unequal in their progress, and during the whole series the organic life of the earth has undergone a corresponding alteration. This alteration also has been gradual, but complete; after a certain interval not a single species existing which had lived at the commencement of the period. This complete renewal of the forms of life also appears to have occurred several times:— That from the last of the geological epochs to the present or historical epoch, the change of organic life has been gradual: the first appearance of animals now existing can in many cases be traced, their numbers gradually increasing in the more recent formations, while other species continually die out and disappear, so that the present condition of the organic world is clearly derived by a natural process of gradual extinction and creation of species from that of the latest geological periods. We may therefore safely infer a like gradation and natural sequence from one geological epoch to another.

Now, taking this as a fair statement of the results of geological inquiry, we see that the present geographical distribution of life upon the earth must be the result of all the previous changes, both of the surface of the earth itself and of its inhabitants. Many causes, no doubt, have operated of which we must ever remain in ignorance, and we may, therefore, expect to find many details very difficult of explanation, and in attempting to give one, must allow ourselves to call into our service geological changes which it is highly probable may have occurred, though we have no direct evidence of their individual operation.

The great increase of our knowledge within the last twenty years, both of the present and past history of the organic world, has accumulated a body of facts which should afford a sufficient foundation for a comprehensive law embracing and explaining them all, and giving a direction to new researches. It is about ten years since the idea of such a law suggested itself to the writer of this essay, and he has since taken every opportunity of testing it by all the newly-ascertained facts with which he has become acquainted, or has been able to observe himself. These have all served to convince him of the correctness of his hypothesis. Fully to enter into such a subject would occupy much space, and it is only in consequence of some views having been lately promulgated, he believes, in a wrong direction, that he now ventures to present his ideas to the public, with only such obvious illustrations of the arguments and results as occur to him in a place far removed from all means of reference and exact information.

A LAW DEDUCED FROM WELL-KNOWN GEOGRAPHICAL AND GEOLOGICAL FACTS

The following propositions in Organic Geography and Geology give the main facts on which the hypothesis is founded.

Geography

1. Large groups, such as classes and orders, are generally spread over the whole earth, while smaller ones, such as fam-

ilies and genera, are frequently confined to one portion, often to a very limited district.

2. In widely distributed families the genera are often limited in range; in widely distributed genera, well marked groups of species are peculiar to each geographical district.

3. When a group is confined to one district, and is rich in species, it is almost invariably the case that the most closely allied species are found in the same locality or in closely adjoining localities, and that therefore the natural sequence of the species by affinity is also geographical.

4. In countries of a similar climate, but separated by a wide sea or lofty mountains, the families, genera and species of the one are often represented by closely allied families, genera and species peculiar to the other.

Geology

5. The distribution of the organic world in time is very similar to its present distribution in space.

6. Most of the larger and some small groups extend through several geological periods.

7. In each period, however, there are peculiar groups, found nowhere else, and extending through one or several formations.

8. Species of one genus, or genera of one family occurring in the same geological time are more closely allied than those separated in time.

9. As generally in geography no species or genus occurs in two very distant localities without being also found in intermediate places, so in geology the life of a species or genus has not been interrupted. In other words, no group or species has come into existence twice.

10. The following law may be deduced from these facts:—
Every species has come into existence coincident both in space and time with a pre-existing closely allied species.

CHARLES DARWIN
(1809–82)

From *The Origin of Species*

RECAPITULATION AND CONCLUSION

As this whole volume is one long agrument, it may be convenient to the reader to have the leading facts and inferences briefly recapitulated.

That many and serious objections may be advanced against the theory of descent with modification through variation and natural selection, I do not deny. I have endeavoured to give to them their full force. Nothing at first can appear more difficult to believe than that the more complex organs and instincts have been perfected, not by means superior to, though analogous with, human reason, but by the accumulation of innumerable slight variations, each good for the individual possessor. Nevertheless, this difficulty, though appearing to our imagination insuperably great, cannot be considered real if we admit the following propositions, namely, that all parts of the organisation and instincts offer, at least, individual differences—that there is a struggle for existence leading to the preservation of profitable deviations of structure or instinct—and, lastly, that gradations in the state of perfection of each organ may have existed, each good of its kind. The truth of these propositions cannot, I think, be disputed.

It is, no doubt, extremely difficult even to conjecture by what gradations many structures have been perfected, more especially amongst broken and failing groups of organic

beings, which have suffered much extinction, but we see so many strange gradations in nature, that we ought to be extremely cautious in saying that any organ or instinct, or any whole structure, could not have arrived at its present state by many graduated steps. There are, it must be admitted, cases of special difficulty opposed to the theory of natural selection; and one of the most curious of these is the existence in the same community of two or three defined castes of workers or sterile female ants; but I have attempted to show how these difficulties can be mastered.

* * *

Turning to geographical distribution, the difficulties encountered on the theory of descent with modification are serious enough. All the individuals of the same species, and all the species of the same genus, or even higher group, are descended from common parents; and therefore, in however distant and isolated parts of the world they may now be found, they must in the course of successive generations have travelled from some one point to all the others. We are often wholly unable even to conjecture how this could have been effected. Yet, as we have reason to believe that some species have retained the same specific form for very long periods of time, immensely long as measured by years, too much stress ought not to be laid on the occasional wide diffusion of the same species; for during very long periods there will always have been a good chance for wide migration by many means. A broken or interrupted range may often be accounted for by the extinction of the species in the intermediate regions. It cannot be denied that we are as yet very ignorant as to the full extent of the various climatal and geographical changes which have affected the earth during modern periods; and such changes will often have facilitated migration. As an example, I have attempted to show how potent has been the influence of the Glacial period on the distribution of the same and of allied species throughout the world. We are as yet profoundly ignorant of the many occasional means of transport. With respect to distinct species of the same genus inhabiting distant and isolated regions, as the process of modification has necessarily been slow, all the means of migration will have been possible during a very long period; and consequently the difficulty of the wide diffusion of the species of the same genus is in some degree lessened.

As according to the theory of natural selection an interminable number of intermediate forms must have existed, linking together all the species in each group by gradations as fine as are our existing varieties, it may be asked: Why do we not see these linking forms all around us? Why are not all organic beings blended together in an inextricable chaos? With respect to existing forms, we should remember that we have no right to expect (excepting in rare cases) to discover *directly* connecting links between them, but only between each and some extinct and supplanted form. Even on a wide area, which has during a long period remained continuous, and of which the climatic and other conditions of life change insensibly in proceeding from a district occupied by one species into another district occupied by a closely allied species, we have no just right to expect often to find intermediate varieties in the intermediate zones. For we have reason to believe that only a few species of a genus ever undergo change; the other species becoming utterly extinct and leaving no modified progeny. Of the species which do change, only a few within the same country change at the same time; and all modifications are slowly effected. I have also shown that the intermediate varieties which probably at first existed in the intermediate zones, would be liable to be supplanted by the allied forms on either hand; for the latter, from existing in greater numbers, would generally be modified and improved at a quicker rate than the intermediate varieties, which existed in lesser numbers; so that the intermediate varieties would, in the long run, be supplanted and exterminated.

On this doctrine of the extermination of an infinitude of connecting links, between the living and extinct inhabitants of the world, and at each successive period between the extinct and still older species, why is not every geological formation charged with such links? Why does not every collection of fossil remains afford plain evidence of the gradation and mutation of the forms of life? Although geological research has undoubtedly revealed the former existence of many links, bringing numerous forms of life much closer together, it does not yield the infinitely many fine gradations between past and present species required on the theory; and this is the most obvious of the many objections which may be urged against it. Why, again, do whole groups of allied species appear, though this appearance is often false, to have come in suddenly on the successive geological stages? Although we now know that organic beings appeared on this globe, at a period incalculably remote, long before the lowest bed of the Cambrian system was deposited, why do we not find beneath this system great piles of strata stored with the remains of the progenitors of the Cambrian fossils? For on the theory, such strata must somewhere have been deposited at these ancient and utterly unknown epochs of the world's history.

I can answer these questions and objections only on the supposition that the geological record is far more imperfect than most geologists believe. The number of specimens in all our museums is absolutely as nothing compared with the countless generations of countless species which have certainly existed. The parent-form of any two or more species would not be in all its characters directly intermediate between its modified offspring, any more than the rock-pigeon is directly intermediate in crop and tail between its descendants, the pouter and fantail pigeons. We should not be able to recognise a species as the parent of another and modified species, if we were to examine the two ever so closely, unless we possessed most of the intermediate links; and owing to the imperfection of the geological record, we have no just right to expect to find so many links. If two or three, or even more linking forms were discovered, they would simply be ranked by many naturalists as so many new species, more especially if found in different geological sub-stages, let their differences be ever so slight. Numerous existing doubtful forms could be named which are probably varieties; but who will pretend that in future ages so many fossil links will be discovered that naturalists will be able to decide whether or not these doubtful forms ought to be called varieties? Only a small portion of the world has been geologically explored. Only organic beings of certain classes can be preserved in a fossil condition, at least in any great number. Many species when once formed never undergo any further change but become extinct without leaving modified descendants; and the periods, during which species have undergone modification, though long as measured by years, have probably been short in comparison with the periods during which they retain the same form. It is the dominant and widely ranging species which vary most fre-

quently and vary most, and varieties are often at first local—
both causes rendering the discovery of intermediate links in
any one formation less likely. Local varieties will not spread
into other and distant regions until they are considerably
modified and improved; and when they have spread, and are
discovered in a geological formation, they appear as if suddenly created there, and will be simply classed as new species. Most formations have been intermittent in their
accumulation; and their duration has probably been shorter
than the average duration of specific forms. Successive formations are in most cases separated from each other by blank
intervals of time of great length; for fossiliferous formations
thick enough to resist future degradations can as a general
rule be accumulated only where much sediment is deposited
on the subsiding bed of the sea. During the alternate periods
of elevation and of stationary level the record will generally be
blank. During these latter periods there will probably be more
variability in the forms of life; during periods of subsidence,
more extinction.

With respect to the absence of strata rich in fossils beneath
the Cambrian formation, I can recur only to the hypothesis
given in the tenth chapter; namely, that though our continents and oceans have endured for an enormous period in
nearly their present relative positions, we have no reason to
assume that this has always been the case; consequently formations much older than any now known may lie buried beneath the great oceans. With respect to the lapse of time not
having been sufficient since our planet was consolidated for
the assumed amount of organic change, and this objection, as
urged by Sir William Thompson,[22] is probably one of the
gravest as yet advanced, I can only say, firstly, that we do not
know at what rate species change as measured by years, and
secondly, that many philosophers are not as yet willing to admit that we know enough of the constitution of the universe
and of the interior of our globe to speculate with safety on its
past duration.

That the geological record is imperfect all will admit; but

[22]Baron Kelvin of Largs (1824–1907), generally regarded as the founder of British
physics.

that it is imperfect to the degree required by our theory, few will be inclined to admit. If we look to long enough intervals of time, geology plainly declares that species have all changed; and they have changed in the manner required by the theory, or they have changed slowly and in a graduated manner. We clearly see this in the fossil remains from consecutive formations invariably being much more closely related to each other, than are the fossils from widely separated formations.

Such is the sum of the several chief objections and difficulties which may be justly urged against the theory; and I have now briefly recapitulated the answers and explanations which, as far as I can see, may be given. I have felt these difficulties far too heavily during many years to doubt their weight. But it deserves especial notice that the more important objections relate to questions on which we are confessedly ignorant; nor do we know how ignorant we are. We do not know all the possible transitional gradations between the simplest and the most perfect organs; it cannot be pretended that we know all the varied means of Distribution during the long lapse of years, or that we know how imperfect is the Geological Record. Serious as these several objections are, in my judgment they are by no means sufficient to overthrow the theory of descent with subsequent modification.

Now let us turn to the other side of the argument. Under domestication we see much variability, caused, or at least excited, by changed conditions of life; but often in so obscure a manner, that we are tempted to consider the variations as spontaneous. Variability is governed by many complex laws,—by correlated growth, compensation, the increased use and disuse of parts, and the definite action of the surrounding conditions. There is much difficulty in ascertaining how largely our domestic productions have been modified; but we may safely infer that the amount has been large, and that modifications can be inherited for long periods. As long as the conditions of life remain the same, we have reason to believe that a modification, which has already been inherited for many generations, may continue to be inherited for an almost infinite number of generations. On the other hand, we have evidence that variability when it has once come into play, does not cease under domestication for a very long period; nor do we know that it ever ceases, for new varieties are still occasionally produced by our oldest domesticated productions.

Variability is not actually caused by man; he only unintentionally exposes organic beings to new conditions of life, and then nature acts on the organisation and causes it to vary. But man can and does select the variations given to him by nature, and thus accumulates them in any desired manner. He thus adapts animals and plants for his own benefit or pleasure. He may do this methodically, or he may do it unconsciously by preserving the individuals most useful or pleasing to him without any intention of altering the breed. It is certain that he can largely influence the character of a breed by selecting, in each successive generation, individual differences so slight as to be inappreciable except by an educated eye. This unconscious process of selection has been the great agency in the formation of the most distinct and useful domestic breeds. That many breeds produced by man have to a large extent the character of natural species, is shown by the inextricable doubts whether many of them are varieties or aboriginally distinct species.

There is no reason why the principles which have acted so efficiently under domestication should not have acted under nature. In the survival of favoured individuals and races, during the constantly-recurrent Struggle for Existence, we see a powerful and ever-acting form of Selection. The struggle for existence inevitably follows from the high geometrical ratio of increase which is common to all organic beings. This high rate of increase is proved by calculation,—by the rapid increase of many animals and plants during a succession of peculiar seasons, and when naturalised in new countries. More individuals are born than can possibly survive. A grain in the balance may determine which individuals shall live and which shall die,—which variety or species shall increase in number, and which shall decrease, or finally become extinct. As the individuals of the same species come in all respects into the closest competition with each other, the struggle will generally be most severe between them; it will be almost equally severe between the varieties of the same species, and next in severity between the species of the same genus. On the other hand the struggle will often be severe between beings remote in the scale of nature. The slightest advantage in certain individuals, at any age or during any season, over those with which they come into competition, or better adaptation in however slight a degree to the surrounding physical conditions, will, in the long run, turn the balance.

With animals having separated sexes, there will be in most cases a struggle between the males for the possession of the females. The most vigorous males, or those which have most successfully struggled with their conditions of life, will generally leave most progeny. But success will often depend on the males having special weapons, or means of defence, or charms; and a slight advantage will lead to victory.

As geology plainly proclaims that each land has undergone great physical changes, we might have expected to find that organic beings have varied under nature, in the same way as they have varied under domestication. And if there has been any variability under nature, it would be an unaccountable fact if natural selection had not come into play. It has often been asserted, but the assertion is incapable of proof, that the amount of variation under nature is a strictly limited quantity. Man, though acting on external characters alone and often capriciously, can produce within a short period a great result by adding up mere individual differences in his domestic productions; and every one admits that species present individual differences. But, besides such differences, all naturalists admit that natural varieties exist, which are considered sufficiently distinct to be worthy of record in systematic works. No one has drawn any clear distinction between individual differences and slight varieties; or between more plainly marked varieties and sub-species, and species. On separate continents, and on different parts of the same continent when divided by barriers of any kind, and on outlying islands, what a multitude of forms exist, which some experienced naturalists rank as varieties, others as geographical races or subspecies, and others as distinct, though closely allied species!

If, then, animals and plants do vary, let it be ever so slightly or slowly, why should not variations or individual differences, which are in any way beneficial, be preserved and accumulated through natural selection, or the survival of the fittest? If man can by patience select variations useful to him, why, under changing and complex conditions of life, should not variations useful to nature's living products often arise, and be preserved or selected? What limit can be put to this power, acting during long ages and rigidly scrutinising the whole constitution, structure, and habits of each creature,— favouring the good and rejecting the bad? I can see no limit to this power, in slowly and beautifully adapting each form to the most complex relations of life. The theory of natural selection, even if we look no farther than this, seems to be in the highest degree probable. I have already recapitulated, as fairly as I could, the opposed difficulties and objections; now let us turn to the special facts and arguments in favour of the theory.

On the view that species are only strongly marked and permanent varieties, and that each species first existed as a variety, we can see why it is that no line of demarcation can be drawn between species, commonly supposed to have been produced by special acts of creation, and varieties which are acknowledged to have been produced by secondary laws. On this same view we can understand how it is that in a region where many species of a genus have been produced, and where they now flourish, these same species should present many varieties; for where the manufactory of species has been active, we might expect, as a general rule, to find it still in action; and this is the case if varieties be incipient species. Moreover, the species of the larger genera, which afford the greater number of varieties or incipient species, retain to a certain degree the character of varieties; for they differ from each other by a less amount of difference than do the species of smaller genera. The closely allied species also of the larger genera apparently have restricted ranges, and in their affinities they are clustered in little groups round other species—in both respects resembling varieties. These are strange relations on the view that each species was independently created, but are intelligible if each existed first as a variety.

As each species tends by its geometrical rate of reproduction to increase inordinately in number; and as the modified descendants of each species will be enabled to increase by as much as they become more diversified in habits and structure, so as to be able to seize on many and widely different places in the economy of nature, there will be a constant tendency in natural selection to preserve the most divergent offspring of any one species. Hence, during a long-continued course of modification, the slight differences characteristic of varieties of the same species, tend to be augmented into the greater differences characteristic of the species of the same genus. New and improved varieties will inevitably supplant and exterminate the older, less improved, and intermediate varieties; and thus species are rendered to a large extent defined and distinct objects. Dominant species belonging to the larger groups within each class tend to give birth to new and dominant forms; so that each large group tends to become still larger, and at the same time more divergent in character. But as all groups cannot thus go on increasing in size, for the world would not hold them, the more dominant groups beat the less dominant. This tendency in the large groups to go on increasing in size and diverging in character, together with the inevitable contingency of much extinction, explains the arrangement of all the forms of life in groups subordinate to groups, all within a few great classes, which has prevailed throughout all time. This grand fact of the grouping of all organic beings under what is called the Natural System, is utterly inexplicable on the theory of creation.

As natural selection acts solely by accumulating slight, successive, favourable variations, it can produce no great or sudden modifications; it can act only by short and slow steps. Hence, the canon of "Natura non facit saltum," which every fresh addition to our knowledge tends to confirm, is on this theory intelligible. We can see why throughout nature the same general end is gained by an almost infinite diversity of means, for every peculiarity when once acquired is long inherited, and structures already modified in many different ways have to be adapted for the same general purpose. We can, in short, see why nature is prodigal in variety, though niggard in innovation. But why this should be a law of nature if each species has been independently created no man can explain.

Many other facts are, as it seems to me, explicable on this theory. How strange it is that a bird, under the form of a woodpecker, should prey on insects on the ground; that upland geese which rarely or never swim, should possess webbed feet; that a thrush-like bird should dive and feed on sub-aquatic insects: and that a petrel should have the habits and structure fitting it for the life of an awk! and so in endless other cases. But on the view of each species constantly trying to increase in number, with natural selection always ready to adapt the slowly varying descendants of each to any unoccupied or ill-occupied place in nature, these facts cease to be strange, or might even have been anticipated.

We can to a certain extent understand how it is that there is so much beauty throughout nature; for this may be largely attributed to the agency of selection. That beauty, according to our sense of it, is not universal, must be admitted by every one who will look at some venomous snakes, at some fishes, and at certain hideous bats with a distorted resemblance to the human face. Sexual selection has given the most brilliant colours, elegant patterns, and other ornaments to the males, and sometimes to both sexes of many birds, butterflies, and other animals. With birds it has often rendered the voice of the male musical to the female, as well as to our ears. Flowers and fruit have been rendered conspicuous by brilliant colours in contrast with the green foliage, in order that the flowers may be readily seen, visited and fertilised by insects, and the seeds disseminated by birds. How it comes that certain colours, sounds, and forms should give pleasure to man and the lower animals,—that is, how the sense of beauty in its simplest form was first acquired,—we do not know any more than how certain odours and flavours were first rendered agreeable.

As natural selection acts by competition, it adapts and improves the inhabitants of each country only in relation to their co-inhabitants; so that we need feel no surprise at the species of any one country, although on the ordinary view supposed to have been created and specially adapted for that country, being beaten and supplanted by the naturalised productions from another land. Nor ought we to marvel if all the contrivances in nature be not, as far as we can judge, absolutely perfect, as in the case even of the human eye; or if some of them be abhorrent to our ideas of fitness. We need not marvel at the sting of the bee, when used against an enemy, causing the bee's own death; at drones being produced in such great numbers for one single act, and being then slaughtered by their sterile sisters; at the astonishing waste of pollen by our fir-trees; at the instinctive hatred of the queen-bee for her own fertile daughters; at the Ichneumonidæ [parasitic flies] feeding within the living bodies of caterpillars; or at other such cases. The wonder indeed is, on the theory of natural selection, that more cases of the want of absolute perfection have not been detected.

The complex and little known laws governing the production of varieties are the same, as far as we can judge, with the laws which have governed the production of distinct species. In both cases physical conditions seem to have produced some direct and definite effect, but how much we cannot say. Thus, when varieties enter any new station, they occasionally assume some of the characters proper to the species of that station. With both varieties and species, use and disuse seem to have produced a considerable effect; for it is impossible to resist this conclusion when we look, for instance, at the logger-headed duck, which has wings incapable of flight, in nearly the same condition as in the domestic duck; or when we look at the burrowing tucu-tucu [rat native to the Strait of Magellan], which is occasionally blind, and then at certain moles, which are habitually blind and have their eyes covered with skin; or when we look at the blind animals inhabiting the dark caves of America and Europe. With varieties and species, correlated variation seems to have played an important part, so that when one part has been modified other parts have been necessarily modified. With both parties and species, reversions to long-lost characters occasionally occur. How inexplicable on the theory of creation is the occasional appearance of stripes on the shoulders and legs of the several species of the horse-genus and of their hybrids! How simply is this fact explained if we believe that these species are all descended from a striped progenitor, in the same manner as the several domestic breeds of the pigeon are descended from the blue and barred rock-pigeon!

On the ordinary view of each species having been independently created, why should specific characters, or those by which the species of the same genus differ from each other, be more variable than generic characters in which they all agree? Why, for instance, should the colour of a flower be more likely to vary in any one species of a genus, if the other species possess differently coloured flowers, than if all possessed the same coloured flowers? If species are only well-marked varieties, of which the characters have become in a high degree permanent, we can understand this fact; for they have already varied since they branched off from a common progenitor in certain characters, by which they have come to be specifically distinct from each other; therefore these same characters would be more likely again to vary than the generic characters which have been inherited without change for an immense period. It is inexplicable on the theory of creation why a part developed in a very unusual manner in one species alone of a genus, and therefore, as we may naturally infer, of great importance to that species, should be eminently liable to variation; but, on our view, this part has undergone, since the several species branched off from a common progenitor, an unusual amount of variability and modification, and therefore we might expect the part generally to be still variable. But a part may be developed in the most unusual manner, like the wing of a bat, and yet not be more variable than any other structure, if the part be common to many subordinate forms, that is, if it has been inherited for a very long period; for in this case, it will have been rendered constant by long-continued natural selection.

Glancing at instincts, marvellous as some are, they offer no greater difficulty than do corporeal structures on the theory of the natural selection of successive slight, but profitable modi-

fications. We can thus understand why nature moves by graduated steps in endowing different animals of the same class
with their several instincts. I have attempted to show how
much light the principle of gradation throws on the admirable
architectural powers of the hive-bee. Habit no doubt often
comes into play in modifying instincts; but it certainly is not
indispensable, as we see in the case of neuter insects, which
leave no progeny to inherit the effects of long-continued
habit. On the view of all the species of the same genus having
descended from a common parent, and having inherited
much in common, we can understand how it is that allied species, when placed under widely different conditions of life,
yet follow nearly the same instincts; why the thrushes of trop-
ical and temperate South America, for instance, line their
nests with mud like our British species. On the view of instincts having been slowly acquired through natural selection,
we need not marvel at some instincts being not perfect and
liable to mistakes, and at many instincts causing other animals
to suffer.

If species be only well-marked and permanent varieties,
we can at once see why their crossed offspring should follow
the same complex laws in their degrees and kinds of resemblance to their parents,—in being absorbed into each
other by successive crosses, and in other such points,—as do
the crossed offspring of acknowledged varieties. This similarity would be a strange fact, if species had been independently created and varieties had been produced through
secondary laws.

If we admit that the geological record is imperfect to an
extreme degree, then the facts, which the record does give,
strongly support the theory of descent with modification.
New species have come on the stage slowly and at successive
intervals; and the amount of change, after equal intervals of
time, is widely different in different groups. The extinction of
species and of whole groups of species which has played so
conspicuous a part in the history of the organic world, almost
inevitably follows from the principle of natural selection; for
old forms are supplanted by new and improved forms. Neither single species nor groups of species reappear when the

chain of ordinary generation is once broken. The gradual diffusion of dominant forms, with the slow modification of their descendants, causes the forms of life, after long intervals of time, to appear as if they had changed simultaneously throughout the world. The fact of the fossil remains of each formation being in some degree intermediate in character between the fossils in the formations above and below, is simply explained by their intermediate position in the chain of descent. The grand fact that all extinct beings can be classed with all recent beings, naturally follows from the living and the extinct being the offspring of common parents. As species have generally diverged in character during their long course of descent and modification, we can understand why it is that the more ancient forms, or early progenitors of each group, so often occupy a position in some degree intermediate between existing groups. Recent forms are generally looked upon as being, on the whole, higher in the scale of organisation than ancient forms; and they must be higher, in so far as the later and more improved forms have conquered the older and less improved forms in the struggle for life; they have also generally had their organs more specialised for different functions. This fact is perfectly compatible with numerous beings still retaining simple and but little improved structures, fitted for simple conditions of life; it is likewise compatible with some forms having retrograded in organisation, by having become at each stage of descent better fitted for new and degraded habits of life. Lastly, the wonderful law of the long endurance of allied forms on the same continent,—of marsupials in Australia, of Edentata [animals without incisor or canine teeth] in America, and other such cases,—is intelligible, for within the same country the existing and the extinct will be closely allied by descent.

*   *   *

I have now recapitulated the facts and considerations which have thoroughly convinced me that species have been modified, during a long course of descent. This has been effected chiefly through the natural selection of numerous successive, slight, favourable variations; aided in an important manner by the inherited effects of the use and disuse of parts; and in an unimportant manner, that is in relation to adaptive structures, whether past or present, by the direct action of external conditions, and by variations which seem to us in our ignorance to arise spontaneously. It appears that I formerly underrated the frequency and value of these latter forms of variation, as leading to permanent modifications of structure independently of natural selection. But as my conclusions have lately been much misrepresented, and it has been stated that I attribute the modification of species exclusively to natural selection, I may be permitted to remark that in the first edition of this work, and subsequently, I placed in a most conspicuous position—namely, at the close of the Introduction—the following words: "I am convinced that natural selection has been the main but not the exclusive means of modification." This has been of no avail. Great is the power of steady misrepresentation; but the history of science shows that fortunately this power does not long endure.

It can hardly be supposed that a false theory would explain, in so satisfactory a manner as does the theory of natural selection, the several large classes of facts above specified. It has recently been objected that this is an unsafe method of arguing; but it is a method used in judging of the common events of life, and has often been used by the greatest natural philosophers. The undulatory theory of light has thus been arrived at; and the belief in the revolution of the earth on its own axis was until lately supported by hardly any direct evidence. It is no valid objection that science as yet throws no light on the far higher problem of the essence or origin of life. Who can explain what is the essence of the attraction of gravity? No one now objects to following out the results consequent on this unknown element of attraction; notwithstanding that Leibnitz[23] formerly accused Newton of introducing "occult qualities and miracles into philosophy."

I see no good reason why the views given in this volume should shock the religious feelings of any one. It is satisfactory, as showing how transient such impressions are, to re-

[23]Gottfried Wilhelm Leibnitz (1646–1716), a philosopher.

member that the greatest discovery ever made by man, namely, the law of the attraction of gravity, was also attacked by Leibnitz, "as subversive of natural, and inferentially of revealed, religion." A celebrated author and divine has written to me that "he has gradually learnt to see that it is just as noble a conception of the Deity to believe that He created a few original forms capable of self-development into other and needful forms, as to believe that He required a fresh act of creation to supply the voids caused by the action of His laws."

Why, it may be asked, until recently did nearly all the most eminent living naturalists and geologists disbelieve in the mutability of species? It cannot be asserted that organic beings in a state of nature are subject to no variation; it cannot be proved that the amount of variation in the course of long ages is a limited quality; no clear distinction has been, or can be, drawn between species and well-marked varieties. It cannot be maintained that species when intercrossed are invariably sterile, and varieties invariably fertile; or that sterility is a special endowment and sign of creation. The belief that species were immutable productions was almost unavoidable as long as the history of the world was thought to be of short duration; and now that we have acquired some idea of the lapse of time, we are too apt to assume, without proof, that the geological record is so perfect that it would have afforded us plain evidence of the mutation of species, if they had undergone mutation.

But the chief cause of our natural unwillingness to admit that one species has given birth to clear and distinct species, is that we are always slow in admitting great changes of which we do not see the steps. The difficulty is the same as that felt by so many geologists, when Lyell first insisted that long lines of inland cliffs had been formed, and great valleys excavated, by the agencies which we see still at work. The mind cannot possibly grasp the full meaning of the term of even a million years; it cannot add up and perceive the full effects of many slight variations, accumulated during an almost infinite number of generations.

*   *   *

I believe that animals are descended from at most only four or five progenitors, and plants from an equal or lesser number.

Analogy would lead me one step farther, namely, to the belief that all animals and plants are descended from some one prototype. But analogy may be a deceitful guide. Nevertheless all living things have much in common, in their chemical composition, their cellular structure, their laws of growth, and their liability to injurious influences. We see this even in so trifling a fact as that the same poison often similarly affects plants and animals; or that the poison secreted by the gallfly produces monstrous growths on the wild rose or oak-tree. With all organic beings excepting perhaps some of the very lowest, sexual production seems to be essentially similar. With all, as far as is at present known, the germinal vesicle is the same; so that all organisms start from a common origin. If we look even to the two main divisions—namely, to the animal and vegetable kingdoms—certain low forms are so far intermediate in character that naturalists have disputed to which kingdom they should be referred. As Professor Asa Gray[24] has remarked, "The spores and other reproductive bodies of many of the lower algae may claim to have first a characteristically animal, and then an unequivocally vegetable existence." Therefore, on the principle of natural selection with divergence of character, it does not seem incredible that, from such low and intermediate form, both animals and plants may have been developed; and, if we admit this, we must likewise admit that all the organic beings which have ever lived on this earth may be descended from some one primordial form. But this inference is chiefly grounded on analogy and it is immaterial whether or not it be accepted. No doubt it is possible, as Mr. G. H. Lewes[25] has urged, that at the first commencement of life many different forms were evolved; but if so we may conclude that only a very few have left modified descendants. For, as I have recently remarked in regard to the members of each great kingdom, such as the

[24]Asa Gray (1810–88), an American naturalist.
[25]G. H. Lewes (1817–1878), author of *Problems of Life and Mind*.

Vertebrata, Articulata, &c., we have distinct evidence in their embryological homologous and rudimentary structures that within each kingdom all the members are descended from a single progenitor.

When the views advanced by me in this volume, and by Mr. Wallace, or when analogous views on the origin of species are generally admitted, we can dimly foresee that there will be a considerable revolution in natural history. Systematists will be able to pursue their labours as at present; but they will not be incessantly haunted by the shadowy doubt whether this or that form be a true species. This, I feel sure and I speak after experience, will be no slight relief. The endless disputes whether or not some fifty species of British brambles are good species will cease. Systematists will have only to decide (not that this will be easy) whether any form be sufficiently constant and distinct from other forms, to be capable of definition; and if definable, whether the differences be sufficiently important to deserve a specific name. This latter point will become a far more essential consideration than it is at present; for differences, however slight, between any two forms, if not blended by intermediate gradations, are looked at by most naturalists as sufficient to raise both forms to the rank of species.

Hereafter we shall be compelled to acknowledge that the only distinction between species and well-marked varieties is, that the latter are known, or believed, to be connected at the present day by intermediate gradations, whereas species were formerly thus connected. Hence, without rejecting the consideration of the present existence of intermediate gradations between any two forms we shall be led to weigh more carefully and to value higher the actual amount of difference between them. It is quite possible that forms now generally acknowledged to be merely varieties may hereafter be thought worthy of specific names; and in this case scientific and common language will come into accordance. In short, we shall have to treat species in the same manner as those naturalists treat genera, who admit that genera are merely artificial combinations made for convenience. This may not be a cheering prospect; but we shall at least be free from the vain search for the undiscovered and undiscoverable essence of the term species.

The other and more general departments of natural history will rise greatly in interest. The terms used by naturalists, of affinity, relationship, community of type, paternity, morphology, adaptive characters, rudimentary and aborted organs, &c., will cease to be metaphorical, and will have a plain signification. When we no longer look at an organic being as a savage looks at a ship, as something wholly beyond his comprehension; when we regard every production of nature as one which has had a long history; when we contemplate every complex structure and instinct as the summing up of many contrivances, each useful to the possessor, in the same way as any great mechanical invention is the summing up of the labour, the experience, the reason, and even the blunders of numerous workmen; when we thus view each organic being, how far more interesting—I speak from experience—does the study of natural history become!

A grand and almost untrodden field of inquiry will be opened, on the causes and laws of variation, on correlation, on the effects of use and disuse, on the direct action of external conditions, and so forth. The study of domestic productions will rise immensely in value. A new variety raised by man will be a more important and interesting subject for study than one more species added to the infinitude of already recorded species. Our classifications will come to be, as far as they can be so made, genealogies; and will then truly give what may be called the plan of creation. The rules for classifying will no doubt become simpler when we have a definite object in view. We possess no pedigrees or armorial bearings; and we have to discover and trace the many diverging lines of descent in our natural genealogies, by characters of any kind which have long been inherited. Rudimentary organs will speak infallibly with respect to the nature of long-lost structures. Species and groups of species which are called aberrant, and which may fancifully be called living fossils, will aid us in forming a picture of the ancient forms of life. Embryology will often reveal to us the structure, in some degree obscured, of the prototype of each great class.

When we feel assured that all the individuals of the same

species, and all the closely allied species of most genera, have within a not very remote period descended from one parent, and have migrated from some one birth-place; and when we better know the many means of migration, then, by the light which geology now throws, and will continue to throw, on former changes of climate and the level of the land, we shall surely be enabled to trace in an admirable manner the former migrations of the inhabitants of the whole world. Even at present, by comparing the differences between the inhabitants of the sea on the opposite sides of a continent, and the nature of the various inhabitants on that continent, in relation to their apparent means of immigration, some light can be thrown on ancient geography.

The noble science of Geology loses glory from the extreme imperfection of the record. The crust of the earth with its imbedded remains must not be looked at as a well-filled museum, but as a poor collection made at hazard and at rare intervals. The accumulation of each great fossiliferous formation will be recognised as having depended on an unusual concurrence of favourable circumstances, and the blank intervals between the successive stages as having been of vast duration. But we shall be able to gauge with some security the duration of these intervals by a comparison of the preceding and succeeding organic forms. We must be cautious in attempting to correlate as strictly contemporaneous two formations, which do not include many identical species, by the general succession of the forms of life. As species are produced and exterminated by slowly acting and still existing causes, and not by miraculous acts of creation; and as the most important of all causes of organic change is one which is almost independent of altered and perhaps suddenly altered physical conditions, namely, the mutual relation of organism to organism,—the improvement of one organism entailing the improvement or the extermination of others; it follows, that the amount of organic change in the fossils of consecutive formations probably serves as a fair measure of the relative though not actual lapse of time. A number of species, however, keeping in a body might remain for a long period unchanged, whilst within the same period several of these species by migrating into new countries and coming into competition with foreign associates, might become modified; so that we must not overrate the accuracy of organic change as a measure of time.

In the future I see open fields for far more important researches. Psychology will be securely based on the foundation already well laid by Mr. Herbert Spencer, that of the necessary acquirement of each mental power and capacity by gradation. Much light will be thrown on the origin of man and his history.

Authors of the highest eminence seem to be fully satisfied with the view that each species has been independently created. To my mind it accords better with what we know of the laws impressed on matter by the Creator, that the production and extinction of the past and present inhabitants of the world should have been due to secondary causes, like those determining the birth and death of the individual. When I view all beings not as special creations, but as the lineal descendants of some few beings which lived long before the first bed of the Cambrian system was deposited, they seem to me to become ennobled. Judging from the past, we may safely infer that not one living species will transmit its unaltered likeness to a distant futurity. And of the species now living very few will transmit progeny of any kind to a far distant futurity; for the manner in which all organic beings are grouped, shows that the greater number of species in each genus, and all the species in many genera, have left no descendants, but have become utterly extinct. We can so far take a prophetic glance into futurity as to foretell that it will be the common and widely-spread species, belonging to the larger and dominant groups within each class, which will ultimately prevail and procreate new and dominant species. As all the living forms of life are the lineal descendants of those which lived long before the Cambrian epoch, we may feel certain that the ordinary succession by generation has never once been broken, and that no cataclysm has desolated the whole world. Hence we may look with some confidence to a secure future of great length. And as natural selection works solely by and for the good of each being, all corporeal and mental endowments will tend to progress towards perfection.

It is interesting to contemplate a tangled bank, clothed with many plants of many kinds, with birds singing on the bushes, with various insects flitting about, and with worms crawling through the damp earth, and to reflect that these elaborately constructed forms, so different from each other, and dependent upon each other in so complex a manner, have all been produced by laws acting around us. These laws, taken in the largest sense, being Growth with Reproduction; Inheritance which is almost implied by reproduction; Variability from the indirect and direct action of the conditions of life and from use and disuse; a Ratio of Increase so high as to lead to a Struggle for Life, and as a consequence to Natural Selection, entailing Divergence of Character and the Extinction of less-improved forms. Thus, from the war of nature, from famine and death, the most exalted object which we are capable of conceiving, namely, the production of the higher animals, directly follows. There is grandeur in this view of life, with its several powers, having been originally breathed by the Creator into a few forms or into one; and that, whilst this planet has gone cycling on according to the fixed law of gravity, from so simple a beginning endless forms most beautiful and most wonderful have been, and are being evolved.

ALFRED, LORD TENNYSON
(1809–92)

From *In Memoriam*

LV

The wish, that of the living whole
    No life may fail beyond the grave,
    Derives it not from what we have
The likest God within the soul?

Are God and Nature then at strife,
    That Nature lends such evil dreams?
    So careful of the type she seems,
So careless of the single life,

That I, considering everywhere
    Her secret meaning in her deeds,
    And finding that of fifty seeds
She often brings but one to bear,

I falter where I firmly trod,
    And falling with my weight of cares
    Upon the great world's altar-stairs
That slope thro' darkness up to God,

I stretch lame hands of faith, and grope,
    And gather dust and chaff, and call
    To what I feel is Lord of all,
And faintly trust the larger hope.

LVI

"So careful of the type," but no.
    From scarped cliff and quarried stone
    She cries, "A thousand types are gone;
I care for nothing, all shall go.

"Thou makest thine appeal to me.
    I bring to life, I bring to death;
    The spirit does but mean the breath:
I know no more," And he, shall he,

Man, her last work, who seem'd so fair,
    Such splendid purpose in his eyes,
    Who roll'd the psalm to wintry skies,
Who built him fanes of fruitless prayer,

Who trusted God was love indeed
    And love Creation's final law—
    Tho' Nature, red in tooth and claw
With ravine, shriek'd against his creed—

Who lov'd, who suffer'd countless ills,
    Who battled for the True, the Just,
    Be blown about the desert dust,
Or seal'd within the iron hills?

No more? A monster then, a dream,
    A discord, Dragons of the prime,
    That tare each other in their slime,
Were mellow music match'd with him,

O life as futile, then, as frail!
    O for thy voice to soothe and bless!
    What hope of answer, or redress?
Behind the veil, behind the veil.

CXVIII

Contemplate all this work of Time,
    The giant laboring in his youth;
    Nor dreams of human love and truth,
As dying Nature's earth and lime;

But trust that those we call the dead
    Are breathers of an ampler day
    For ever nobler ends. They say,
The solid earth whereon we tread

In tracts of fluent heat began,
    And grew to seeming-random forms,
    The seeming prey of cyclic storms,
Till at the last arose the Man;

Who throve and branch'd from clime to clime,
  The herald of a higher race,
  And of himself in higher place,
If so he type this work of time

Within himself, from more to more;
  Or, crown'd with attributes of woe
  Like glories, move his course, and show
That life is not as idle ore,

But iron dug from central gloom,
  And heated hot with burning fears,
  And dipt in baths of hissing tears,
And batter'd with the shocks of doom

To shape and use. Arise and fly
  The reeling Faun, the sensual feast;
  Move upward, working out the beast,
And let the ape and tiger die.

BY AN EVOLUTIONIST

The Lord let the house of a brute to the soul of a man,
  And the man said, "Am I your debtor?"
And the Lord—"Not yet: but make it as clean as you can,
  And then I will let you a better."

1

If my body come from brutes, my soul uncertain, or a fable,
  Why not bask amid the senses while the sun of morning
    shines,
I, the finer brute rejoicing in my hounds, and in my stable,
  Youth and health, and birth and wealth, and choice of
    women and of wines?

2

What has thou done for me, grim Old Age, save breaking my
    bones on the rack?

Would I had passed in the morning that looks so bright
from afar!

Old Age

Done for thee? starved the wild beast that was linked with
thee eighty years back.
Less weight now for the ladder-of-heaven that hangs on a
star.

1

If my body come from brutes, though somewhat finer than
their own.
I am heir, and this my kingdom. Shall the royal voice be
mute?
No, but if the rebel subject seek to drag me from the throne,
Hold the scepter, Human Soul, and rule thy province of the
brute.

2

I have climbed to the snows of Age, and I gaze at a field in the
Past,
Where I sank with the body at times in the sloughs of a low
desire.
But I hear no yelp of the beast, and the Man is quiet at last
As he stands on the heights of his life with a glimpse of a
height that is higher.

SAMUEL BUTLER

(1835–1902)

From *Erewhon*

THE BOOK OF THE MACHINES, Chapter 23

The writer commences: "There was a time when the earth was to all appearance utterly destitute both of animal and vegetable life, and when according to the opinion of our best philosophers it was simply a hot round ball with a crust gradually cooling. Now if a human being had existed while the earth was in this state and had been allowed to see it as though it were some other world with which he had no concern, and if at the same time he were entirely ignorant of all physical science, would he not have pronounced it impossible that creatures possessed of anything like consciousness should be evolved from the seeming cinder which he was beholding? Would he not have denied that it contained any potentiality of consciousness? Yet in the course of time consciousness came. Is it not possible then that there may be even yet new channels dug out for consciousness, though we can detect no signs of them at present?

"Again. Consciousness, in anything like the present acceptation of the term, having been once a new thing—a thing, as far as we can see, subsequent even to an individual center of action and to a reproductive system (which we see existing in plants without apparent consciousness)—why may not there arise some new phase of mind which shall be as different from all present known phases, as the mind of animals is from that of vegetables?

"It would be absurd to attempt to define such a mental state (or whatever it may be called), inasmuch as it must be something so foreign to man that his experience can give him no help towards conceiving its nature; but surely when we reflect upon the manifold phases of life and consciousness which have been evolved already, it would be rash to say that no others can be developed, and that animal life is the end of all things. There was a time when fire was the end of all things: another when rocks and water were so."

The writer, after enlarging on the above for several pages, proceeded to inquire whether traces of the approach of such a new phase of life could be perceived at present; whether we could see any tenements preparing which might in a remote futurity be adapted for it; whether, in fact, the primordial cell of such a kind of life could be now detected upon earth. In the course of his work he answered this question in the affirmative and pointed to the higher machines.

"There is no security"—to quote his own words—"against the ultimate development of mechanical consciousness, in the fact of machines possessing little consciousness now. A mollusc has not much consciousness. Reflect upon the extraordinary advance which machines have made during the last few hundred years, and note how slowly the animal and vegetable kingdoms are advancing. The more highly organised machines are creatures not so much of yesterday, as of the last five minutes, so to speak, in comparison with past time. Assume for the sake of argument that conscious beings have existed for some twenty million years: see what strides machines have made in the last thousand! May not the world last twenty million years longer? If so, what will they not in the end become? Is it not safer to nip the mischief in the bud and to forbid them further progress?

"But who can say that the vapour engine has not a kind of consciousness? Where does consciousness begin, and where end? Who can draw the line? Who can draw any line? Is not everything interwoven with everything? Is not machinery linked with animal life in an infinite variety of ways? The shell of a hen's egg is made of a delicate white ware and is a machine as much as an egg-cup is: the shell is a device for holding the egg, as much as the egg-cup for holding the shell: both are phases of the same function; the hen makes the shell in her inside, but it is pure pottery. She makes her nest outside of herself for convenience' sake, but the nest is not more of a machine than the egg-shell is. A 'machine' is only a 'device.'"

Then returning to consciousness, and endeavoring to detect its earliest manifestations, the writer continued:

"There is a kind of plant that eats organic food with its flowers: when a fly settles upon the blossom, the petals close upon it and hold it fast till the plant has absorbed the insect into its system; but they will close on nothing but what is good to eat; of a drop of rain or a piece of stick they will take no notice. Curious! that so unconscious a thing should have such a keen eye to its own interest. If this is unconsciousness, where is the use of consciousness?

"Shall we say that the plant does not know what it is doing merely because it has no eyes, or ears, or brains? If we say that it acts mechanically, and mechanically only, shall we not be forced to admit that sundry other and apparently very deliberate actions are also mechanical? If it seems to us that the plant kills and eats a fly mechanically, may it not seem to the plant that a man must kill and eat a sheep mechanically?

"But it may be said that the plant is void of reason, because the growth of a plant is involuntary growth. Given earth, air, and due temperature, the plant must grow: it is like a clock, which being once wound up will go till it is stopped or run down: it is like the wind blowing on the sails of a ship—the ship must go when the wind blows it. But can a healthy boy help growing if he have good meat and drink and clothing? can anything help going as long as it is wound up, or go on after it is run down? Is there not a winding up process everywhere?

"Even a potato in a dark cellar has a certain low cunning about him which serves him in excellent stead. He knows perfectly well what he wants and how to get it. He sees the light coming from the cellar window and sends his shoots crawling straight thereto: they will crawl along the floor and up the wall and out at the cellar window; if there be a little earth anywhere on the journey he will find it and use it for his own ends. What deliberation he may exercise in the matter of his roots when he is planted in the earth is a thing unknown to us, but we can imagine him saying, 'I will have a tuber here and a tuber there, and I will suck whatsoever advantage I can from all my surroundings. This neighbour I will overshadow, and that I will undermine; and what I can do shall be the limit of what I will do. He that is stronger and better placed then I shall overcome me, and he that is weaker I will overcome.'

"The potato says these things by doing them, which is the best of languages. What is consciousness if this is not consciousness? We find it difficult to sympathize with the emotions of a potato; so we do with those of an oyster. Neither of these things makes a noise on being boiled or opened, and noise appeals to us more strongly than anything else, because we make so much about our own sufferings. Since, then, they do not annoy us by any expression of pain we call them emotionless; and so *qua* mankind they are; but mankind is not everybody.

"If it be urged that the action of the potato is chemical and mechanical only, and that it is due to the chemical and mechanical effects of light and heat, the answer would seem to lie in an inquiry whether every sensation is not chemical and mechanical in its operation? whether those things which we deem most purely spiritual are anything but disturbances of equilibrium in an infinite series of levers, beginning with those that are too small for microscopic detection, and going up to the human arm and the appliances which it makes use of? whether there be not a molecular action of thought, whence a dynamical theory of the passions shall be deducible? Whether strictly speaking we should not ask what kind of levers a man is made of rather than what is his temperament? How are they balanced? How much of such and such will it take to weigh them down so as to make him do so and so?"

The writer went on to say that he anticipated a time when it would be possible, by examining a single hair with a powerful microscope, to know whether its owner could be insulted with impunity. He then became more and more obscure, so that I was obliged to give up all attempt at translation; neither did I follow the drift of his argument. On coming to the next part which I could construe, I found that he had changed his ground.

"Either," he proceeds, "a great deal of action that has been called purely mechanical and unconscious must be admitted to contain more elements of consciousness than has been allowed hitherto (and in this case germs of consciousness will be found in many actions of the higher machines)—or (assuming the theory of evolution but at the same time denying the consciousness of vegetable and crystalline action) the race of man has descended from things which had no consciousness at all. In this case there is no *a priori* improbability in the descent of conscious (and more than concious) machines from those which now exist, except that which is suggested by the apparent absence of anything like a reproductive system in the mechanical kingdom. This absence however is only apparent, as I shall presently show.

"Do not let me be misunderstood as living in fear of any actually existing machine; there is probably no known machine which is more than a prototype of future mechanical life. The present machines are to the future as the early Saurians to man. The largest of them will probably greatly diminish in size. Some of the lowest vertebrata attained a much greater bulk than has descended to their more highly organized living representatives, and in like manner a diminution in the size of machines has often attended their development and progress.

"Take the watch, for example; examine its beautiful structure; observe the intelligent play of the minute members which compose it: yet this little creature is but a development of the cumbrous clocks that preceded it; it is no deterioration from them. A day may come when clocks, which certainly at the present time are not diminishing in bulk, will be superseded owing to the universal use of watches, in which case they will become as extinct as ichthyosauri, while the watch, whose tendency has for some years been to decrease in size rather than the contrary, will remain the only existing type of an extinct race.

JOHN UPDIKE
(1932—)

The Naked Ape

**(following, Perhaps All Too Closely, Desmond Morris's
Anthropological Revelations)**

The dinosaur died, and small
    Insectivores (how gruesome!) crawled
From bush to tree, from bug to bud,
    From spider-diet to forest fruit and nut,
Developing bioptic vision and
    The grasping hand.

These perfect monkeys then were faced
    With shrinking groves; the challenged race,
De-Edenized by glacial whim,
    Sent forth from its arboreal cradle him
Who engineered himself to run
    With deer and lion—

The "naked ape." Why naked? Well,
    Upon those meaty plains, that *veldt*
Of prey, as pellmell they competed
    With cheetahs, hairy primates overheated;
Selection pressure, just though cruel,
    Favored the cool.

Unlikeliest of hunters, nude
    And weak and tardy to mature,
This ill-cast carnivore attacked,

With weapons he invented, *in a pack.*
The tribe was born. To set men free,
  The family

Evolved; monogamy occurred.
  The female—sexually alert
Throughout the month, equipped to have
  Pronounced orgasms—perpetrated love.
The married state decreed its *lex*
  Privata: sex.

And Nature, pandering, bestowed
  On virgin ears erotic lobes
And hung on women hemispheres
  That imitate their once-attractive rears:
A social animal disarms
  With frontal charms.

All too erogenous, the ape
  To give his lusts a decent shape
Conceived the cocktail party where
  Unmates refuse to touch each other's hair
And to make small "grooming" talk instead
  Of going to bed.

He drowns his body scents in baths
  And if, in some conflux of paths,
He bumps another, says, "Excuse
  Me, *please.*" He suffers rashes and subdues
Aggressiveness by making fists
  And laundry lists,

Suspension bridges, aeroplanes,
  And charts that show biweekly gains
And losses. Noble animal!
  To try to lead on this terrestrial ball,
With grasping hand and saucy wife,
  The upright life.

LOUIS PASTEUR
(1822–95)

From *The Germ Theory and Its Applications to Medicine and Surgery*

The Sciences gain by mutual support. When, as the result of my first communications on the fermentations in 1857–1858, it appeared that the ferments, properly so-called, are living beings, that the germs of microscopic organisms abound in the surface of all objects, in the air and in water; that the theory of spontaneous generation is chimerical; that wines, beer, vinegar, the blood, urine and all the fluids of the body undergo none of their usual changes in pure air, both Medicine and Surgery received fresh stimulation. A French physician, Dr. Davaine,[26] was fortunate in making the first application of these principles to Medicine, in 1863.

Our researches of the last year, left the etiology of the putrid disease, or septicemia, in a much less advanced condition than that of anthrax. We had demonstrated the probability that septicemia depends upon the presence and growth of a microscopic body, but the absolute proof of this important conclusion was not reached. To demonstrate experimentally that a microscopic organism actually is the cause of a disease and the agent of contagion, I know no other way, in the present state of Science, than to subject the *microbe* (the new and happy term introduced by M. Sedillot[27]) to the method of

[26] Casimir Joseph Davaine (1812–1882), pioneer microbiologist.
[27] Charles Emmanuel Sedillot (1804–1883), director of the school for military medicine.

cultivation out of the body. It may be noted that in twelve successive cultures, each one of only ten cubic centimeters volume, the original drop will be diluted as if placed in a volume of fluid equal to the total volume of the earth. It is just this form of test to which M. Joubert and I subjected the anthrax bacteridium. Having cultivated it a great number of times in a sterile fluid, each culture being started with a minute drop from the preceding, we then demonstrated that the product of the last culture was capable of further development and of acting in the animal tissues by producing anthrax with all its symptoms. Such is—as we believe—the indisputable proof that *anthrax is a bacterial disease.*

Our researches concerning the septic vibrio had not so far been convincing, and it was to fill up this gap that we resumed our experiments. To this end, we attempted the cultivation of the septic vibrio from an animal dead of septicemia. It is worth noting that all of our first experiments failed, despite the variety of culture media we employed— urine, beer yeast water, meat water, etc. Our culture media were not sterile, but we found—most commonly—a microscopic organism showing no relationship to the septic vibrio, and presenting the form, common enough elsewhere, of chains of extremely minute spherical granules possessed of no virulence whatever.[28] This was an impurity, introduced, unknown to us, at the same time as the septic vibrio; and the germ undoubtedly passed from the intestines—always inflamed and distended in septicemic animals—into the abdominal fluids from which we took our original cultures of the septic vibrio. If this explanation of the contamination of our cultures was correct, we ought to find a pure culture of the septic vibrio in the heart's blood of an animal recently dead of septicemia. This was what happened, but a new difficulty presented itself; all our cultures remained sterile. Furthermore this sterility was accompanied by loss in the culture media of (the original) virulence.

It occurred to us that the septic vibrio might be an obligatory anaërobe and that the sterility of our inoculated culture

---

[28] Possibly septicemic streptococci.

fluids might be due to the destruction of the septic vibrio by the atmospheric oxygen dissolved in the fluids. The Academy may remember that I have previously demonstrated facts of this nature in regard to the vibrio of butyric fermentation, which not only lives without air but is killed by the air.

It was necessary therefore to attempt to cultivate the septic vibrio either in a vacuum or in the presence of inert gases— such as carbonic acid.

Results justified our attempt; the septic vibrio grew easily in a complete vacuum, and no less easily in the presence of pure carbonic acid.

These results have a necessary corollary. If a fluid containing septic vibrios be exposed to pure air, the vibrios should be killed and all virulence should disappear. This is actually the case. If some drops of septic serum be spread horizontally in a tube and in a very thin layer, the fluid will become absolutely harmless in less than half a day, even if at first it was so virulent as to produce death upon the inoculation of the smallest portion of a drop.

Furthermore all the vibrios, which crowded the liquid as motile threads, are destroyed and disappear. After the action of the air, only fine amorphous granules can be found, unfit for culture as well as for the transmission of any disease whatever. It might be said that the air burned the vibrios.

If it is a terrifying thought that life is at the mercy of the multiplication of these minute bodies, it is a consoling hope that Science will not always remain powerless before such enemies, since for example at the very beginning of the study we find that simple exposure to air is sufficient at times to destroy them.

JOHN UPDIKE
(1932—)

Amoeba

Mindless, meaning no harm,
it ingested me.
It moved on silent pseudopods
to where I was born, inert, and I
was inside.

Digestive acids burned my skin.
Enzymes nuzzled knees and eyes.
My ego like a conjugated verb
retained its root, a narrow fear
of being qualified.

Alas, suffixes swarmed.
I lost my mother's arms, my teeth,
my laugh, my protruding faith;
Reduced to the O of a final sigh,
in time I died.

MURIEL RUKEYSER
(1913—80)

The Conjugation of the Paramecium

This has nothing
to do with
propagating

The species
is continued
as so many are
(among the smaller creatures)
by fission

(and this species
is very small
next in order to
the amoeba, the beginning one)

The paramecium
achieves, then,
immortality
by dividing

But when
the paramecium
desires renewal
strength   another joy
this is what
the paramecium does:

The paramecium
lies down beside
another
paramecium

Slowly   inexplicably
the exchange
takes place
in which
some bits
of the nucleus of each
are exchanged

for some bits
of the nucleus
of the other

This is called
the conjugation of the paramecium.

RICHARD EBERHART

(1904—)

The Cancer Cells

Today I saw a picture of the cancer cells,
Sinister shapes with menacing attitudes.
They had outgrown their test-tube and advanced,
Sinister shapes with menacing attitudes,
Into a world beyond, a virulent laughing gang.
They looked like art itself, like the artist's mind,
Powerful shaker, and the taker of new forms.

Some are revulsed to see these spiky shapes;
It is the world of the future too come to.
Nothing could be more vivid than their language,
Lethal, sparkling and irregular stars,
The murderous design of the universe,
The hectic dance of the passionate cancer cells.
O just phenomena to the calculating eye,
Originals of imagination. I flew
With them in a piled exuberance of time,
My own malignance in their racy, beautiful gestures
Quick and lean: and in their riot too
I saw the stance of the artist's make,
The fixed form in the massive fluxion.

I think Leonardo would have in his disinterest
Enjoyed them precisely with a sharp pencil.

MIROSLAV HOLUB

(1923—)

Wings

We have
a microscopic anatomy
of the whale
this
gives
Man
assurance
—William Carlos Williams

We have
a map of the universe
for microbes,

we have
a map of a microbe
for the universe.

We have
a Grand Master of chess
made of electronic circuits.

But above all
we have
the ability
to sort peas,
to cup water in our hands,
to seek
the right screw
under the sofa
for hours

This
gives us
wings.

NATHANIEL TARN

(1928—)

From *The Beautiful Contradictions*

FIVE

Looking into the eyes of babies in experiments
born without the normal pressure on their skulls
thinking they are going to put an end to philosophy
when some development of this begins to breed monsters

and that the chase through probability of the genius
the great kick he gives through his mother as he comes out
the clarity of the air surrounding him later in life
however much his body might take revenge on him
his mind crack between the diameter of his skull and the crown

                    reality comprises

that the immeasurable heave of the whole race
to bring this animal to the tree's crest and enthrone him there
may be gone forever in a moment of medical history
like the passing of some art   or an old migration
of all the birds together in the arms of the same wind
the way the planet used to turn in one direction with one purpose

                    frightens a lot

I remember on the shores of the most beautiful lake in the world
whose name in its own language means abundance of waters
as if the volcanos surrounding it had broken open the earth
there in the village of Saint James of Compostela one cold night
not the cereus-scented summer nights in which a voice I never
        traced
sang those heartbreaking serenades to no one known
a visiting couple gave birth in the market place
the father gnawing the cord like a rat to free the child
and before leaving in the morning they were given the
                    freedom of the place

                    I mean the child was given

MAY SWENSON

(1919—)

The DNA Molecule

THE DNA MOLECULE
THE DNA MOLECULE
THE DNA MOLECULE
is The Nude Descending a Staircase
a circular one.
See the undersurfaces
of the spiral
treads and
the spaces
in between.
She is descending and at the same time
ascending and she moves around herself. For
she is the staircase "a protoplasmic framework
an internal scaffolding
that twists and turns."

She is a double helix mounting and dismounting
around the swivel of her imaginary spine. The Nude
named DNA can be constructed as a model with matches and
a ribbon of tape. Be sure to use only 4 colors on 2 white
strands of twistable tape. "Only matches of complementary
colors may be placed opposite each other. The pairs
are to be red and green and yellow and blue."
Make your model as high as the Empire
State Building and you have an acceptable
replica of The Nude.
But and this is harder you must make her move
in a continuous coil
an alpha helix a double spiral
downward and upward at once
and you must make her increase while at the same
time occupying the same field.
She must be made "to maintain a basic topography"
changing yet remaining stable

if she is to perform her function which is to produce
and reproduce the microsphere.
Such a sphere is invisible to but omnipresent
in the naked eye of The Nude.
It contains "a central region and an outer membrane"
making it able to divide "to make exact copies of
itself without limit."

The Nude has "the capacity for
replication and transcription" of
all genesis. She ingests and
regurgitates the genetic material
it being the material of her own
cell-self. From single she becomes
double and from double single.
As a woman ingests the demon sperm and with the same membrane
regurgitates the mitotic double of herself upon the

slide of time so the DNA
MOLECULE produces with a little
pop at the waistline of its viscous drop
a new microsphere the same size
as herself which proceeds singly to grow
in order to divide and double itself.
So from single to double and double to single and

mounting while descending she
expands while contracts she proliferates while
disappearing at both of her ends.

Remember that red can only be opposite green
and blue opposite yellow. Remember that the
complementary pairs of matches must differ slightly in
length "for nature's pairs can be made only with units
whose structures permit an interplay of forces
between the partners."

I fixed a blue match opposite a red
match of the same length
in defiance of the rules pointed them
away from the center on the double-stranded
tape. I saw laid a number of eggs

on eggs on the sticky side of a twig.
I saw a worm with many feet grow out
of an egg.

The worm climbed the twig a single helix and gobbled
the magnified edge of a leaf
in quick enormous bites.

It then secreted out of itself a gray floss
with which it wrapped itself tail first and so on
until it had completely muffled
and encased itself head last as in a mummy pouch.

I saw plushy iridescent wings push
moistly out of the pouch. At first glued
together they began to part. On each wing

I saw a large blue eye
open forever in the expression of resurrection.
The new Nude released the flanges
of her wings
stretching herself to touch

at all points
the outermost rim
of the noösphere.

I saw that for her body from which the
wings expanded
she had retained
the worm.

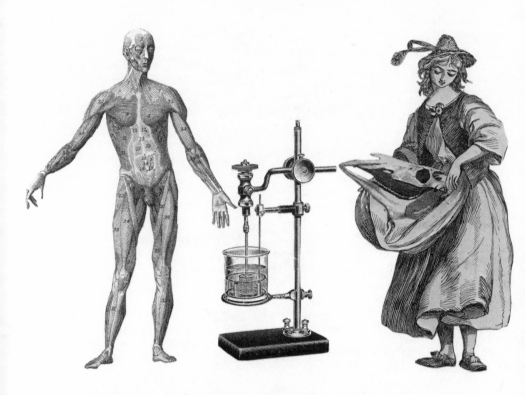

KATHLEEN RAINE
(1908—)

The Human Form Divine

The human contours are so easily lost.
Only close your eyes and you seem a forest
Of dense vegetation, and the lurking beast

That in the night springs from the cover
Tears with tiger's mouth your living creatures,
A thousand innocent victims without name that suffer.

Science applies its insect-lenses to the form divine
As up the red river (all life comes from the sea)
Swim strange monsters, amoeboid erythrean spawn.

Rock-face of bone, alluvium of cartilage
Remote from man as the surface of the moon
Are vast and unexplored interior desert ranges,

And autonomous cells
Grow like unreaped fields of waving corn.
Air filters through the lungs' fine branches as through trees.

Chemistry dissolves the goddess in the alembic,
Venus the white queen, the universal matrix,
Down to molecular hexagons and carbon-chains,

And the male nerve-impulse, monition of reality,
Conveys the charge, dynamic of non-entity
That sparks across the void *ex nihilo*.

At the extreme of consciousness, prayer
Fixes hands and feet immobile to a chair,
Transmutes all heaven and earth into a globe of air,

And soul streams away out of the top of the head
Like flame in a lamp-glass carried in the draught
Of the celestial fire kindled in the solar plexus.

Oh man, oh Garden of Eden, there is nothing
But the will of love to uphold your seeming world,
To trace in chaos the contours of your beloved form!

ZBIGNIEW HERBERT
(1924—)

Biology Teacher

I can't remember
his face

he stood high above me
on long spread legs
I saw
the little gold chain
the ash-grey frock coat
and the thin neck
on which was pinned
a dead necktie

he was the first to show us
the leg of a dead frog
which touched by a needle
violently contracts

he led us
through a golden microscope
to the intimate life
of our great-grandfather
the slipper animalcule

he brought
a dark kernel
and said: claviceps

encouraged by him
I became a father
at the age of ten
when after tense anticipation
a yellow sprout appeared
from a chestnut submerged in water
and everything broke into song
all around

in the second year of the war
the rascals of history
killed the teacher of biology

if he reached heaven—

perhaps he is walking now
on the long rays
dressed in grey stockings
with a huge net
and a green box
gaily swinging on his back

but if he didn't go up there—

when on a forest path
I meet a beetle scrambling
up a hill of sand
I come close to him
click my heels
and say:
—Good morning professor
would you let me help you

I lift him over carefully
and for a long time look after him
until he disappears
in the dark faculty room
at the end of the corridor of leaves

KENNETH REXROTH

(1905–82)

Survival of the Fittest

I realize as I
Cast out over the lake
At thirteen thousand feet—
I don't know where you are.
It has been years since we
Married and had children
By people neither of
Us knew in the old days.
But I still catch fish with flies
Made from your blonde pubic hair.

INDEX

A